Understanding Development

Developmental biology is seemingly well understood, with development widely accepted as being a series of programmed changes through which an egg turns into an adult organism, or a seed matures into a plant. However, the picture is much more complex than that: is it all genetically controlled or does environment have an influence? Is the final adult stage the target of development and everything else just a build-up to that point? Are developmental strategies the same in plants as in animals? How do we consider development in single-celled organisms? In this concise, engaging volume, Alessandro Minelli, a leading developmental biologist, addresses these key questions. Using familiar examples and easy-to-follow arguments, he offers fresh alternatives to a number of preconceptions and stereotypes, awakening the reader to the disparity of developmental phenomena across all main branches of the tree of life.

Alessandro Minelli is a former professor of zoology and, in retirement, a senior scientist at the University of Padova, Italy. He is an honorary fellow of the Royal Entomological Society and the Italian Society for Developmental and Cell Biology. He is the author of numerous books on evolutionary and developmental biology, including *The Development of Animal Form* (Cambridge, 2003), *Plant Evolutionary Developmental Biology* (Cambridge, 2018) and *The Biology of Reproduction* (Cambridge, 2019; with Giuseppe Fusco).

The ***Understanding Life*** series is for anyone wanting an engaging and concise way into a key biological topic. Offering a multidisciplinary perspective, these accessible guides address common misconceptions and misunderstandings in a thoughtful way to help stimulate debate and encourage a more in-depth understanding. Written by leading thinkers in each field, these books are for anyone wanting an expert overview that will enable clearer thinking on each topic.

Series Editor: Kostas Kampourakis http://kampourakis.com

Published titles:

Understanding Evolution	Kostas Kampourakis	9781108746083
Understanding Coronavirus	Raul Rabadan	9781108826716
Understanding Development	Alessandro Minelli	9781108799232
Understanding Evo-Devo	Wallace Arthur	9781108819466

Forthcoming:

Understanding Genes	Kostas Kampourakis	9781108812825
Understanding DNA Ancestry	Sheldon Krimsky	9781108816038
Understanding Intelligence	Ken Richardson	9781108940368
Understanding Metaphors in the Life Sciences	Andrew S. Reynolds	9781108940498
Understanding Creationism	Glenn Branch	9781108927505
Understanding Species	John S. Wilkins	9781108987196
Understanding the Nature–Nurture Debate	Eric Turkheimer	9781108958165
Understanding How Science Explains the World	Kevin McCain	9781108995504
Understanding Cancer	Robin Hesketh	9781009005999
Understanding Forensic DNA	Suzanne Bell and John Butler	9781009044011
Understanding Race	Rob DeSalle and Ian Tattersall	9781009055581
Understanding Fertility	Gab Kovacs	9781009054164

Understanding Development

ALESSANDRO MINELLI
University of Padova

CAMBRIDGE
UNIVERSITY PRESS

477 Williamstown Road, Port Melbourne, VIC 3207, Australia

314–321, 3rd Floor, Plot 3, Splendor Forum, Jasola District Centre,
New Delhi – 110025, India

79 Anson Road, #06–04/06, Singapore 079906

Cambridge University Press is part of the University of Cambridge.

It furthers the University's mission by disseminating knowledge in the pursuit of education, learning and research at the highest international levels of excellence.

www.cambridge.org
Information on this title: www.cambridge.org/9781108836777
DOI: 10.1017/9781108872287

First published 2021

Printed in the United Kingdom by TJ Books Limited, Padstow Cornwall

A catalogue record for this publication is available from the British Library.

ISBN 978-1-108-83677-7 Hardback
ISBN 978-1-108-79923-2 Paperback

'This is the finest book on the principles underpinning biological development that I have read in a long time. It is succinct, thoughtful and full of examples, offering wise reflection on the diversity of developmental phenomena across the whole tree of life. Understanding Development is especially notable for its organization into 48 sections comprising 8 chapters. Each section subtitle states a key lesson to be learned through brief historical and theoretical expositions, well-chosen examples, and stories of odd-ball and familiar life forms. Every lesson overturns some conventional wisdom or common knowledge that cannot stand up to the wondrous diversity of life on Earth. Minelli's broad, deep knowledge of the field is expressed with an engaging contrarian spirit that serves his larger goal: to prompt a reassessment of the state of contemporary understanding of development in a way accessible to novice and expert alike.'

James Griesemer, University of California–Davis, USA

'Developmental biology has been described as the process by which a fertilized egg is transformed into a multicellular organism. But is it? In this thoughtful and erudite book, Alessandro Minelli forces us to step back and reconsider the subject. Using an astonishing range of examples, from pythons to lichens and from sponges to ciliates, Minelli challenges a series of generalizations and preconceptions. We see how development is not only the process of building adults, why development does not have end-points, how development need not start with a fertilized egg, why we must be careful with the concept of developmental genes, and much more. After reading this book, you might not think about developmental biology in the same way again.'

Peter Holland, University of Oxford, UK

'Developmental biology is a highly dynamic area of the life sciences, and it also lacks a unifying theoretical framework and must rely on general principles derived from a small number of well-studied model organisms. In Understanding Development, Minelli channels an encyclopaedic knowledge of biological diversity to show convincingly the need for a more expansive concept of development that can embrace the variability and complexity of life. Minelli surveys the interplay of generalizations and exceptions that arise in the study of development, tracing out important open conceptual challenges facing researchers today. Engagingly written and always insightful, this book is highly recommended to biologists, philosophers of biology, and historians interested in grappling with a fundamental and active problem area in the contemporary landscape of biological thought.'

James DiFrisco, KU Leuven, Belgium

To the inspiring disparity of Life

Contents

Foreword

Alessandro Minelli has written an exceptionally rich book with great insights. *Understanding Development* shows that in contrast to our adult-centric and anthropocentric view of development, there is a variety of developmental processes in nature. The author effectively debunks numerous misunderstandings about development, some of which you may never have thought of before. The whole book is structured in such a way that all misunderstandings are explicitly discussed and addressed one after the other. In doing so, the author provides exceptionally clear examples from a variety of organisms, which clearly show the complexities of developmental processes and his exceptional knowledge of the topic. Minelli takes readers on a delightful and informative voyage across all forms of life and shows that development can be quite different from what we know from our own experience. He effectively makes the case that to understand life we need to look at other forms of life and their developmental processes. The present book is a fantastic means for doing this; once you have read it, you will feel stunned by the unity and diversity of life that are presented throughout.

Kostas Kampourakis, Series Editor

Preface

For centuries, the study of development was strictly descriptive, and the main tool available to the researcher was the microscope.

At the beginning of the nineteenth century, this discipline witnessed one of its greatest successes: the discovery of the mammalian egg by Karl Ernst von Baer. Shortly thereafter, the cell theory enunciated by Theodor Schwann in 1839 consolidated life sciences by recognizing a similar organization in animals and plants: cells, the structural elements of which multicellular organisms are made, are also the building blocks of development.

Experimental embryology emerged towards the end of the century. Observation was now complemented by mechanical manipulation and by exposure of eggs and embryos to a diversity of chemicals. The resulting discoveries impinged strongly on the interpretation of developmental phenomena. However, the comparative spirit that had hitherto pushed embryologists to conduct their research on a sample of species as numerous and varied as possible began to vanish. Biological research now focused increasingly on a few model species – from Gregor Mendel's peas to Thomas H. Morgan's fruit flies, from Wilhelm Roux's frogs to Hans Driesch's sea urchins, from Theodor Boveri's large roundworms to Hans Spemann's newts.

The naturalistic season of developmental biology seemed to be definitely over, except for its solid legacy in the notion of development as the story of the changes that transform an egg into an adult animal, or a seed into a mature tree.

In the meantime, advances in cell biology fuelled increasingly successful studies of the mechanisms responsible for these changes, from cell proliferation to growth to differentiation. Some of these processes are seemingly common to all organisms, at least to multicellular organisms; others are more specific, for example to plants or to animals only, or even to smaller groups, but it seems reasonable to ignore the minor differences that may exist between one species and another, and instead to emphasize patterns and mechanisms shared by the widest range of organisms. These results became the increasingly solid corpus of general biology.

A decisive turning point was dictated by the entry into the molecular era. The study of genes was no longer limited to the rules of transmission from one generation to the next, but opened up to the investigation of where and when genes are expressed (that is, transcribed into messenger RNA molecules and eventually translated into proteins) during development, and how their expression is regulated. It did not take long to discover that many processes depend on the expression of the same genes in animals as diverse as mice and fruit flies. The temptation to extrapolate general principles from experiments restricted to a few model species was stronger than ever. But before long, many researchers realized that the taxonomic sampling of the living world must be widened again.

From time to time, biologists (and philosophers of biology) feel the need to revisit the basic principles of developmental biology, but this discipline turns out to have only a modest and uncertain body of theory, especially if compared to evolutionary biology.

One might expect that the list of contentious conceptual issues in developmental biology would shrink with the continuous progress of experimental research but, on the contrary, this list becomes longer and longer. Many of these critical issues must be left in the hands of skilled professionals; others, however, are based on traditional misconceptions that can be addressed here.

There are many excellent books, both popular and academic, that describe the stages of embryonic development of animals, the expression of genes involved in patterning a flower or the trunk of a fly, the molecular dialogues between the cells, or the intricate regulatory networks whose protagonists are the genes and their products.

This book is different: it is full of stories involving quite a number of different plants and animals, fungi and protists (single-celled organisms). The track will not be dictated by taxonomic subdivision, but by the problems to be faced to lighten our vision of development from a long list of preconceptions and unjustified generalizations, unfortunately shared by a number of professionals.

Life is a product of history. This cannot be ignored in the fields of developmental and evolutionary biology, the scope of which is the study of change. Therefore, developmental biology cannot omit a systematic exploration of the many different forms in which development takes place in the different groups of living beings.

In Chapter 1, in addition to providing a historical framework, I discuss a possible definition of development and the need to abandon the finalism still latent in developmental biology today. In later chapters I discuss cells (Chapter 2), embryos (Chapter 4), developmental sequences (Chapter 5) and genes (Chapter 6). I do this not to summarize the notions that modern textbooks present with all the necessary technical detail, but rather to address, at each level, the most serious generalizations. The remaining chapters are dedicated to aspects that, in different ways, also affect the philosophy of biology. I discuss here individuals (Chapter 3), regularity of form (Chapter 7) and developmental ecology, with several pages dedicated to temporal aspects such as age, senescence, and the articulation of individual development into a sequence of steps (Chapter 8).

In 1802, Treviranus and Lamarck introduced, independently, the name 'biology' for the science of living beings. More than 200 years later, the time seems to have come to approach biology as the science of all life forms and to avoid reducing it to abstract generalizations.

In the following pages there are rather more stories about animals than about plants, fungi and other kinds of organisms. In part, this is justified by the amazing wealth and complexity of developmental patterns and processes exhibited by animals; in part it is the consequence of my professional rooting in zoological disciplines. This bias notwithstanding, I hope that the reader will share my fascination with the inspiring disparity of developmental mechanisms behind the 'endless forms most beautiful' that evolved along all branches of the tree of life.

Acknowledgements

I am very grateful to Katrina Halliday and Kostas Kampourakis for inviting me to write this book and for assisting me throughout the whole editorial process; also sincerely acknowledged are the contributions of Olivia Boult and Samuel Fearnley at Cambridge University Press. Lindsay Nightingale has efficiently removed my many barbarisms through a very sensible and constructive copy-editing process. A large number of valuable suggestions on a previous version of the text were provided by Wallace Arthur, Giuseppe Fusco, Kostas Kampourakis and Stefano Tiozzo; useful comments have also been provided by Ivan Amato. My wife Maria Pia offered valuable help with the illustrations.

1 Defining Development, if Possible

Development of What?

Development Is Not Necessarily the History of the Individual

At the beginning of our exploration of developmental phenomena, it seems reasonable to address a semantic question: what do we mean by development? Let us focus on the development of living organisms, without worrying about what development may mean, for example, to an economist or an educator.

What can be considered as development is a controversial issue. A few years ago, a group of biologists and philosophers of biology thought it necessary to consider this question seriously. Overall, the debate involved 24 scholars. Two important things emerged from their responses. First, only half of those concerned said that a definition of development was necessary; the others argued that they could safely do without, and one even added that a definition of development is impossible. Second, the proposed definitions were very different from one another, to the point that several important biological phenomena would fall within the sphere of developmental biology for some scholars but not for others.

A look at the list of proposed definitions is useful. It will serve as a guide for our itinerary, not so much to seek answers to our questions as to widen horizons as much as possible and to try to formulate sensible questions. Here are the definitions as proposed. Development is…

- the process by which a single cell gives rise to a complex multicellular organism;

- the generation of a new individual form;
- the change of biological form over time;
- the temporal change of organization along the life cycle;
- the biological reading of DNA-encoded gene networks that determine the structure of the organism;
- the irreversible increase in the complexity of a biological system over time.

In different form, all of these definitions capture important aspects of development and many of the problems on which most research focuses. However, it is sensible that from the very first pages of this book, the reader should assume a critical attitude towards a number of positions that it is all too easy to take for granted, e.g. that

- development is a series of structural changes affecting multicellular organisms; single cells per se, including unicellular organisms, do not develop;
- development is about the biological individual – indeed, it is the process that shapes the individual;
- development necessarily entails an increase in complexity;
- the body contains (multiple copies of) a programme according to which it is formed;
- development is irreversible.

These widespread beliefs are based, at best, on a generalization of conditions that apply only to some living organisms, not to all. They overemphasize aspects that cover only a part of the phenomena that deserve to be called developmental processes.

An excellent starting point to clear the path through these problems is a few sentences by the great French physiologist Claude Bernard:

> . . . all morphological change is contained in the previous state. This work is pure repetition. [. . .] there is no morphology without predecessors. In reality we do not witness the birth of a new being: we only see a periodic continuation [. . .]. Things happen this way because the being is in a way imprisoned in a series of conditions from which it cannot escape, since they are always repeated in the same way internally and externally.

This text is from the *Leçons sur les phénomènes de la vie communs aux animaux et aux végétaux* (Lectures on the Phenomena of Life Common to Animals and Plants) published in 1878, immediately after Bernard's death.

Thus, in Bernard's vision, development is not the history of the individual from zygote to adult, but a series of changes in which each step depends strongly on the conditions in which the living organism was until then. It is like a chess game, where different choices (some more advantageous for the player, some less so, but this is not important) are generally possible with each move; these choices depend on the current arrangement of the pieces on the board and this, in turn, depends on the previous moves. The comparison between the succession of changes in development and the configurations of the pieces on a chessboard, however, is only valid as long as the game is in full swing. When the game is over, the board is emptied: there is no continuity between one game and the next, whereas there is between one biological generation and the next. It is precisely here that it is advisable to take a closer look, to make clear what should be considered as development.

In fact, we are faced with two possibilities. If we listen to Bernard, development is a process (better, perhaps, a set of processes) that continues through the generations. If instead we follow the popular notion (shared by many professionals), development is the individual story of the changes that transform an egg first into an embryo, then into a juvenile and finally into an adult.

However, there is a way to overcome this dichotomy. If we do not want to leave a good number of important topics outside the scope of developmental biology, we should accept a notion of development consistent with Bernard's observation. The chapters of developmental biology, therefore, will not be (only) those that correspond to moments in the history of an individual, such as cleavage (the division of the egg into a progressively increasing number of cells, the blastomeres) or gastrulation (the embryonic developmental phase shared by almost all animals, during which germ layers are formed; more on germ layers on p. 78). Instead, the mechanisms of regeneration will fall within the scope of this discipline, even if most of what we know about them are the responses to challenges an animal would never face in nature; or the structural changes resulting from a pathological situation, primarily cancer, or those induced by the presence of a parasite.

In this broader perspective, the history of the individual does not become less interesting or less central in developmental biology, except that the succession of stages from egg to adult animal (or from seed to mature tree) is no longer *the* development, but *a particular history* of development. In the pages of this book, there will be space both for individual stories (often very different from what we might describe as the normal development of an individual of our own species) and for developmental processes as such, over a time span that may be shorter or longer than one generation.

Debunking Adultocentrism

Development Is Not Necessarily the Sequence of Changes from Egg to Adult

When we describe an animal or a plant as a 'monster', this is because it departs significantly – although sometimes in one feature only – from the morphology of a typical individual of the species. A calf should have one head rather than two; a fruit fly should have two wings rather than four, as found in a well-known mutant. However, it is not always easy to say what the typical structure of an animal should be.

First, an animal may undergo metamorphosis over the course of its life. A newt, for example, spends many weeks as a tadpole, and the differences between tadpole and post-metamorphic newt are important: the tadpole is an aquatic animal that breathes through gills, while the adult can move on land and breathes through lungs (and skin).

A tadpole and a newt are the same animal; nevertheless, when we identify in the adult morphology the form typical for the species, we give the adult form an absolute value. The egg and the embryonic stages, to continue with larva and juveniles, are thus downgraded to mere preparatory stages. The 'true form' of the newt is the form of the adult. This is acceptable in the everyday use of the term 'newt', but not if we want to understand developmental biology.

As in other situations (I will give more examples in this book), useful suggestions come from the study of individuals that have undergone a less than normal development. Some newts, for example, become sexually mature without having undergone metamorphosis: they retain their larval shape and continue to increase in size, while their gonads mature as in a normal adult. If we follow the standard terminology, these newts, although able to reproduce, are not 'real' adults. The 'true form' of the animal is another.

Second, there are organisms in which it is impossible to recognize a standard form. The tiny fungus *Candida albicans*, for example, can easily switch between a single-cell form, comparable to a yeast, and a filamentous, multicellular one (see p. 27).

When discussing development, it is critically important, even if difficult, to move away from the traditional attitude that deserves the name of adultocentrism, according to which all the embryonic, larval and juvenile stages – and the developmental processes in which they are involved – are only steps or means required to become an adult. This old attitude has not changed much with the modern concept of development as the deployment of a genetic programme, because the latter is intended as a programme for the production of an adult.

In the traditional adultocentric view of development:

- The adult condition is the goal to be reached. However, we will see that this is not always true; moreover, the very notion of adult is sometimes problematic.
- Once the adult condition is reached, development is stopped. If life extends beyond the reproductive stage, the adult faces ageing – a phenomenon that traditionally belongs to the discipline of pathology rather than to the biology of development. But we will see that changes in the organism in post-reproductive age occur according to processes of the same nature as those that characterize previous stages.
- Developmental mechanisms have been consolidated through natural selection, therefore they are adaptive. But we will see that from the point of view of the cellular or molecular mechanisms involved, 'normal' developmental processes and phenomena such as the production of tumours are not necessarily very different.
- The sequence of events that characterize an individual's development is irreversible. On this topic too we will have something to say.

It is difficult to deny that the adultocentric vision contains a good deal of finalism. From this point of view, a comparison between developmental biology and evolutionary biology can be interesting. In the latter, finalism survives only in rather superficial popular versions of the theory, in which evolution is considered synonymous with progress, rather than a continuous and always imperfect adjustment to the changing conditions faced by a population. In developmental biology, however, sentences with a finalistic flavour often come from the pen of authoritative scientists. For example, Eric

Davidson, a scholar to whom we owe major achievements in the molecular genetics of development, wrote that "development is the execution of the genetic programme for the construction of a given species of organism," and that "a particular function of embryonic cells is to interact in specific ways, in order to generate morphological structure."

Also adultocentric is the term 'set-aside cells'. This designates groups of cells found in the larvae of many invertebrates, which are not parts of the larva's organs but remain dormant until metamorphosis. Only at this point, while the larval structures are reabsorbed or lost, does the adult take shape precisely from those cells that, until then, had been 'set-aside', almost with the intention – one might say – of using them later in adult morphogenesis. It would be preferable to say that those cells, rather than set-aside, were temporarily marginalized from active life.

An adultocentric view of development requires that each phase be compatible with the next. In my view, the opposite perspective is much more reasonable: that is, development can proceed so far as each phase is compatible with the previous one. In this perspective, there is no difficulty in including in developmental biology the individual stories that stop before the adult condition is reached. I am referring not just to the philosophically uninteresting case of a developmental history truncated by accident, but to stories in which, through intrinsic causes, development is arrested in a condition other than the normal: the so-called 'monsters'.

Disregarding those created in the lab (often invaluable for the progress of developmental biology), monsters sometimes show up in nature, even in our species. Their study is the subject of a specific scientific discipline, teratology. To approach this field, I suggest we turn the pages of the first treatise on comparative teratology (three volumes of text plus one of plates), published in the years 1832–37 by Isidore Geoffroy Saint-Hilaire. In this work, monsters are arranged according to a classification similar to Linnaeus' distribution of animal and plant species. This exercise is very important: if monsters can be classified, this means that their deviations from the normal condition are not arbitrary, but fall within a finite, perhaps small, number of kinds. And even monsters usually obey the laws of biological form, including two-headed calves or fruit flies with the antennae replaced by two legs, at least in so far as they do not depart from bilateral symmetry.

Growth Trajectories

There Is Not Always a Species-Specific Limit to Individual Growth

In 1864, a year before succeeding his father William as the director of Kew Gardens – one of the most prestigious botanical institutions in the world – Joseph Dalton Hooker, one of Charles Darwin's closest friends, described under the name of *Welwitschia mirabilis* a truly unusual plant discovered 5 years previously by the Austrian botanist Friedrich Welwitsch. The homeland of this unique plant is the desert that extends along the border between what are today Namibia and Angola. Its massive woody trunk, which has no branches, resembles a low stump a few tens of centimetres high. From its upper margin sprout two broad ribbon-shaped leaves, each of which can be up to 4 metres long. The tip, which is the oldest part of the leaf, is dry and frayed, especially in older plants. But the two leaves continue to grow, thanks to the proliferative activity of basal cells, throughout the life of the plant. Specimens a thousand years old are not uncommon, and some are believed to be twice as old.

Welwitschia mirabilis is the only living species of a lineage of gymnosperms – a plant with some affinity with conifers, but not very close to them. In the other major group of seed plants, the angiosperms (flowering plants), there are also a few species with continuously growing leaves: in this case, however, growth takes place from the distal tip, and the whole leaf will wither within a few years. These plants are tropical trees of the mahogany family (Meliaceae), more precisely those classified in the genera *Guarea* and *Chisocheton*.

Indeterminate growth, however, seems to be a widespread feature in trees even if, sooner or later, the process will necessarily come to an end. We will discuss in Chapter 8 whether ageing affects all living beings, or whether some organisms do not experience it. But we do not need to invoke ageing here: even the most robust tree ends up succumbing to attacks by fungi or insects, helped perhaps by severe atmospheric events.

We might think that things are different in animals. In humans, growth in height eventually slows down, then ceases altogether. Other familiar vertebrates follow the same trend. But it would not be safe to generalize. Even among mammals there are species in which growth continues throughout life, even if this slows with the onset of maturity. Examples are bison, giraffes and elephants. There are many more examples of animals with indeterminate growth among the amphibians and, more conspicuously, among fishes such as the grouper. There is also no shortage of examples among invertebrates, for example the giant clams of the genus *Tridacna*, which can live over a century, reaching enormous size. The largest known shell of a giant clam weighs 330 kilograms; the mollusc that produced it weighed perhaps another 20 kilograms.

In other cases, the arrival of reproductive maturity puts a neat end to growth, even if this was previously very rapid. Among the plants, bamboos reach the most extraordinary growth rates, up to 90 centimetres in a single day, but the plants die after their only flowering season. Animal embryos often elongate particularly fast, especially those supplied with a large amount of yolk. The increase in size of a tumour is also often very fast. In the context of normal post-embryonic development, extraordinarily rapid growth is exhibited by many tapeworms. Within 2 weeks after infection, *Hymenolepis diminuta*, a tapeworm 20 to 60 centimetres in length that lives in the rat intestine, increases 3400 times in length and 1.8 million times in weight, producing the fantastic figure of 2200 proglottids (the technical term for the 'segments' of tapeworms).

But there are also animals that go through periods of negative growth. This is not simply a matter of weight loss due to lack of nourishment for a prolonged period, but of a somewhat 'regulated', although regressive, developmental process that allows the animal to resume positive growth when environmental conditions or availability of food are back to normal. Cases of negative growth have been observed in many invertebrate groups, but we will take a look at just three examples.

Under fasting conditions, 1-centimetre-long planarians (a group of free-living flatworm, the most popular of which live in freshwater) can be reduced to a tiny worm less than a millimetre long, but their complex anatomical structure remains substantially preserved, through a proportional reduction of the various organs.

Even more intense is the effect of negative growth in some nemertines, a group of worm-like animals, almost all marine, also known as the ribbon worms. Some nemertines, for example some species of the genus *Lineus*, can endure fasting for more than a year, reducing their size from a few tens of centimetres to a microscopic mass of a few hundred cells: in the process, the gonads, digestive tract and other organs are resorbed. The outcome of this negative growth can be described as a return to an embryonic level of morphological complexity.

The final example is from insects. In conditions of prolonged fasting, the larva of the small beetle *Trogoderma glabrum* progressively decreases in weight and also in length, but it cannot be said that its development is suspended. On the contrary, it goes through a higher number of moults than normal. Whereas under normal feeding conditions the larva of this insect goes through five or six stages, a fasting larva continues to moult an indeterminate number of times, even for years: after each moult, its size is a little smaller.

Uncertain Boundaries

Developmental Change Is Not Necessarily Different from Metabolism

In a part of the tree of life where the divide between unicellular and multicellular conditions is particularly thin, we find a species of tiny marine organisms known as *Salpingoeca rosetta*. This is the best studied species among the choanoflagellates, a small group of microscopic eukaryotes with some characteristics that suggest affinity with the evolutionary lineage of the animals. The trait that gives them their name is the presence of a whip-like structure called a flagellum, surrounded by a showy collar. What is remarkable, of course, is not the flagellum: flagellated cells are present in many groups of organisms, such as the spermatozoa of most animals, including humans. It is the collar that is not as common. This is a sort of circular palisade formed by microvilli, those slender rod-shaped extensions found, for example, on the side of the cells of our digestive tract that faces the lumen of the intestine, greatly increasing their surface area and therefore the efficiency of the absorption of digested substances. The cells of the gut mucosa, however, possess microvilli but not flagella. Cells similar to those of choanoflagellates are characteristic of sponges, where they are called choanocytes.

In both cases (choanoflagellates and choanocytes), the continuous movement of the flagellum and the presence of the surrounding palisade of microvilli allow for efficient circulation of water. This forwards tiny nutrient particles to the cells, which then engulf them. What makes the similarity between choanoflagellates and sponge choanocytes more interesting is that some choanoflagellates, including *Salpingoeca*, often live in groups: that is, they exhibit a rudimentary form of multicellularity. In recent years, DNA comparisons have confirmed that choanoflagellates are indeed among the closest relatives of animals. This has motivated a search in choanoflagellates for clues about the transition from the uni- to multicellular condition: for example, the presence in choanoflagellates of those molecules that allow cells resulting from a mitosis (the normal cell division that produces two daughter cells, both of them with a complete copy of the parent cell's genome) to remain together. A number of these molecules have actually been found in choanoflagellates. In the latter, however, the transition between the single-cell and the multicell organization is not an obligate developmental step, as in animals, but depends on environmental conditions. What triggers this transition? In the case of *Salpingoeca rosetta*, the stimulus is the presence of certain species of bacteria, in particular *Algoriphagus machipongonensis*. But the bacteria do not only act as the trigger of this important morphogenetic event: *Algoriphagus* is also a prey species for the choanoflagellates.

We are thus faced with a situation where one external agent serves both as food and as the trigger of a structural change – that is, it is involved both in metabolism and development. However, this is not an isolated case. Perhaps this distinction was not yet clear before the chain of transformations that we call development took on an autonomous and precise form in the evolutionary lineage of animals. Problematic situations are found almost everywhere.

Let us take snakes as an example. For a large python, coping with food needs is no small problem. Like almost all snakes, pythons are predators, and their appetite can be satisfied only by rather large prey. A python has remarkable strength, and the tightness of its turns allows it to kill a variety of prey, including monkeys and pigs. However, by the time the snake opens its jaws and starts swallowing the prey, its problems have only just begun. The victim is not chewed, cut into small pieces and suitably sprayed with saliva before being pushed down through the oesophagus; instead, it ends up in the stomach still almost untouched. Digestion, which is long and difficult, requires support from all the organs of the snake. In 2 days from the time the python swallowed the prey, the length of the intestinal villi increases fivefold and the muscle mass of the heart ventricle by 40%. This latter increase is due not to cell proliferation but to growth in the amount of contractile proteins in the fibres that make up the heart mass. But this is only the tip of the iceberg. In both pythons and rattlesnakes, in conditions of prolonged fasting, the intestine undergoes morphological and functional regression; after a meal, the gut epithelium resumes its organization and functionality, not only through an increase in cell volume, but also thanks to reactivated cell proliferation. Therefore, the entire cycle of structural and physiological changes that accompany a period of fasting and the subsequent feeding phase translates into alternation between a phase of reduction and dedifferentiation and the following regenerative phase.

The next story confirms that the usual divide between development and metabolism as distinct chapters of biology is subjective. The green sea slug *Elysia chlorotica*, 2 to 3 centimetres long, is common in salt marshes and coastal pools along the eastern coasts of the United States and Canada. The species owes its specific name (*chlorotica*) to its lively green colour. This is due to the filamentous alga *Vaucheria litorea* which it feeds on without digesting it completely. After piercing the cell wall of the alga with the teeth of the scraper (technically, the radula) in its mouth, a sea slug sucks up the contents. In the mollusc's digestive tract, which has extensive ramifications throughout the animal's body, the chloroplasts of the alga remain intact for months, engulfed within the cells of the intestinal wall, and continue to perform photosynthesis. The sugars thus produced contribute to meeting *Elysia*'s food needs: we are therefore in the sphere of metabolism. But this story also affects development closely. The mollusc does not only change from a heterotrophic to an optional autotrophic condition; when it begins feeding on *Vaucheria*, its larva also finds it easier to complete metamorphosis.

Embryos, from Classic Embryology to Modern Developmental Biology

Developmental Biology Is Not the Same as Embryology

In 1673, the Italian physician, physiologist and anatomist Marcello Malpighi published a small work entitled *De formatione pulli in ovo* (On the Chick's Formation in the Egg), a true milestone in the history of embryology. The theme was not new: since Aristotle's time, the chicken had been the most accessible animal in which to study embryonic development. Even though chick embryos develop inside an opaque shell, their study has many advantages, such as availability in large numbers and, more important, the possibility of creating a rigorous time series, by incubating eggs in uniform environmental conditions and measuring the time elapsed between an egg's deposition and the moment in which its shell is broken by a scientist to observe the embryo.

In times closer to Malpighi's, other scholars such as the Italian Girolamo Fabrizi d'Acquapendente and the Englishman William Harvey had also dealt with the chick embryo. However, these authors had made their observations with the naked eye: Malpighi was the first to use a microscope. Without this tool, he would not have been able to make the observations he describes and illustrates in the 25 figures of the booklet. The chick remained the main study object of some of the great researchers of the following century, such as Albrecht von Haller, Lazzaro Spallanzani and Caspar Friedrich Wolff, and of the first half of the following century, such as Christian Pander. Karl Ernst von Baer tackled a more difficult topic, the study of the embryonic development of mammals, culminating in 1827 with his discovery of the elusive egg (or ovum) of these animals. I will look later (p. 70) into von Baer's great work of comparative embryology (two volumes, published respectively in 1828 and 1837), which represents the culmination of these studies.

In the following decades, studies extended to the embryonic development of many different animals, including a number of marine invertebrates, whose transparent eggs and embryos allow much easier observation.

Towards the end of the nineteenth century, purely descriptive study started to be complemented by experimental embryology, which we will deal with later (p. 60). The effects of manipulations (at first exclusively mechanical, later by means of chemical treatments as well), carried out at different developmental times, allowed the identification of critical developmental steps and turning points. Further advances would have to await developments in molecular biology. Scientists would eventually learn to interfere with the expression of those genes for which a morphogenetic effect is known or suspected.

Box 1.1 Animals and Plants: The Life Cycle

The developmental biology flourishing in our days is very different from the descriptive and experimental embryology of the past; however, the reader may profit from a short summary of traditional notions. The brief notes that follow describe the most frequent situations, but there are many deviations from these, several of which are discussed in this book.

In sexually reproducing animals, an individual's development begins with an egg (generally, but not always, a fertilized egg, or zygote). By repeated division of the egg, an embryo arises, consisting of an increasing number of increasingly smaller cells, the blastomeres. This developmental phase is called *cleavage*. In many animals, the embryo goes through a series of characteristic stages called the *morula* (blastomeres packed to form a compact cluster), *blastula* (blastomeres arranged in a single surface layer) and *gastrula* (a sack-shaped embryo, with only one opening, the blastopore or primitive mouth, which leads into the archenteron or primitive intestine). In addition to the two germ layers that make up the gastrula (one external, the *ectoderm*, and one internal, the *endoderm*), in most animals an intermediate germ layer, the *mesoderm*, will also differentiate. Developmental biologists use this verb to describe the processes by which a cell, tissue or body part becomes recognizably different in structure and function from the surrounding cells, tissues or body parts. During the whole embryonic development the animal relies on the nutritional resources (*yolk*) stored in the egg; as a rule, it will be able to feed autonomously at the beginning of its post-embryonic life. At this time, many animals are broadly similar to the future adult; others are very different, developing first as a *larva* that will later metamorphose to *adult*.

An animal's biological cycle typically involves the production of gametes by the sexually mature individual. The union of a male gamete and a female gamete gives rise to a zygote, with which the biological cycle begins anew.

During each biological cycle, an event (*meiosis*) occurs which gives rise to cells with the basic (haploid) number of chromosomes typical of the species (for example, 23 in humans) and an event (*fertilization*) through which the diploid condition (46 chromosomes in our species) is reconstituted by fusion of two haploid cells.

In animals, the haploid phase is limited to gametes. In plants, as described in the text (p. 50), the haploid phase is more conspicuous, especially in ferns and even more in mosses, but to some extent also in flowering plants (p. 59). Here the diploid phase (the *sporophyte*, that is, the plant with leaves, flowers and so on) clearly prevails over the haploid phase. The latter is the *gametophyte* – the pollen grain (male gametophyte, three cells) or the egg+embryo sac complex (female gametophyte, usually made up of seven cells, one of which has two nuclei). The zygote results from the fusion of the egg with one of the cells of a pollen grain.

At each transition from descriptive embryology to experimental embryology to developmental genetics, the understanding of developmental processes has risen to levels unimaginable in the previous stage. This progress, however, has been achieved at a price we are only just starting to realize. For obvious practical reasons, the number and diversity of species studied has tended to shrink. This opens an entire chapter of the life sciences – the biology of model organisms – that will be the subject of the next section.

Up to this point, we have focused mostly on animals. Until the eighteenth century, there were no important contributions to the knowledge of plant developmental biology. Even the existence of sexuality in plants was not accepted before the last years of the seventeenth century. In the middle of the following century, the eminent figure of the German scholar Caspar Friedrich Wolff emerged, author of a theory of generation based on a comparative study of embryonic development in both plant and animal species.

In plant science, the use of the term *embryo* was occasional and lacked a clear circumscription until 1788, when the German botanist Joseph Gaertner successfully used it in the first volume of his large treatise *De fructibus et seminibus plantarum* (*On Plant Fruits and Seeds*). Gaertner defined the embryo as "the most noble and essential part of the seed, the only part that provides the new plant and to which all the other parts of the seed are added for at least temporary use." Until then, there had been no name for the future seedling, in the phase in which it is still enclosed within the seed casings (integuments). It is understandable that Gaertner, faced with the uncertain nomenclature of the few accounts on the seeds of plants published before then, none of which were accurate and comprehensive, turned to animal embryology, which was undoubtedly more advanced. From this literature Gaertner borrowed many terms. 'Embryo' had long been in use to indicate an early stage of development, both of viviparous animals, including our species, and of oviparous ones, such as the chick. Of the many other terms of zoological origin used by Gaertner, some (such as placenta and cotyledon) come from the embryology of viviparous animals, others (such as egg white and yolk) from the embryology of oviparous ones. More than two centuries later, some of these terms have remained in use for both plants and animals, but nobody today would venture to say, for example, that the placenta or cotyledons of plants are the same thing ('homologous structures', to use the technical term) as the animal parts known under the same term.

Unfortunately, the idea of an equivalence between what is called an embryo in either kingdom is widespread even among professionals. To realize this, it is often necessary to lift the veil of apparent modernity provided by questions formulated in molecular terms. For example, some scientific papers address the question of whether there are similarities between the trend of gene expression along the different embryonic stages of animals and plants, but fail to explain how plant and animal developmental phases can actually be compared.

Success and Problems: Studying Development in Model Species

The Success of Modern Developmental Biology Has Not All Been Based on Model Species

What are the main model organisms used in laboratories all over the world for experimental research – the species on which is based much of what we know about development biology, and biology in general?

Among the first entries in the list, we find a few familiar animals, such as mice (*Mus musculus*) and the fruit fly (*Drosophila melanogaster*). We may not pay much attention to fruit flies as they wander among glasses of new wine or overripe fruits that are beginning to rot. However, the role they have played in genetics is well known, and this insect has been increasingly popular since 1908, when the American geneticist Thomas Hunt Morgan and his collaborators began experimenting in what has remained famous as the Fly Room at Columbia University, New York. Other model animals are less popular, but in elementary biology courses it is likely that one will at least encounter photos of zebrafish (*Danio rerio*) and of the tiny nematode worm whose scientific name is *Caenorhabditis elegans*. Among plants, the most fashionable model species is the humble thale cress (*Arabidopsis thaliana*), but also important are tomato (*Solanum lycopersicum*), snapdragon (*Antirrhinum majus*) and rice (*Oryza sativa*), plus the moss *Aphanorrhegma patens* (usually known as *Physcomitrella patens*). Let's add to this list two representatives of the fungi (baker's yeast *Saccharomyces cerevisiae* and the red bread mould *Neurospora crassa*), and the cellular slime mould *Dictyostelium discoideum*, an odd kind of organism whose life cycle will be the subject of a later section (p. 36). Discoveries based on most of these species will be mentioned repeatedly in later chapters. The problem with model species is that the results of observations and experiments limited to them cannot be extrapolated to other species so broadly and uncritically as we might hope.

The small nematode worm *Caenorhabditis elegans*, about a millimetre long, has been reared for research since the middle of the last century, but it was only in 1963 that an important research effort based on this species started, involving a large number of developmental biology labs. Several reasons motivated this choice. In part, these are the usual reasons that make an animal a good candidate for a model species: small size, short life cycle, easy and cheap breeding in the lab. To this, we can add the remarkable transparency of the body of this worm; and, finally, a trait that made *C. elegans* particularly promising for development biologists: in small nematodes and some other invertebrates of similar size, the number of cells is the same in all individuals, both the whole body count and the number of cells in each organ. This characteristic is known as eutely. To be more precise, numbers in the case of *C. elegans* are given for nuclei rather than cells, because some sets of cells are fused to form a multinucleate unit with common cytoplasm – a syncytium (more on this subject on p. 36). In adult hermaphrodite individuals there are 959 nuclei; of the total, 302 belong to neurons, 95 to muscle cells in the body wall and so on. Another 131 cells were formed in the embryo, but underwent programmed cell death (apoptosis; p. 24).

All these cells stem from the fertilized egg, through a series of divisions whose course is rigidly fixed within the species. The highly conserved sequence of mitoses was expected to be associated with a progressive restriction of the final fate of the cell, eventually allocating it to specific parts of the body. The surprise came when it was discovered that the cell's final fate is not fully dictated by its position in the cell lineage tree but also by its interactions with other cells (p. 63).

Drosophila melanogaster is far from being representative of insects in general, precisely because of a quality that is otherwise appreciated in a model species: the short duration of its life cycle. In the lab, at 25 °C, the whole embryonic development of a fruit fly takes just one day. But this speed is the effect of dramatic deviations from the usual insect developmental pathways. In *Drosophila*, the first 13 cycles of mitosis occur in syncytial rather than cellular conditions; a bit later, all body segments will be formed synchronously rather than by adding new segments at the posterior end. Correspondingly, the expression of genes with major effects on establishing the general structure of the embryo is quite different from the spatial and temporal patterns recorded in most other insects. Many ground-breaking discoveries in developmental genetics were made while experimenting with *Drosophila*: a lot of adjustments became necessary in order to apply these results to other insects and non-insect animals.

Problems are no less important in the case of *Arabidopsis thaliana*. It would be unfair to describe this as 'typical', not only of the whole world of flowering plants – about 300 000 species, including herbs, grasses, bushes and trees – but even of the 3700 species in the cabbage family (Brassicaceae) with which it is classified.

The genome of *A. thaliana* is much smaller than most other plant genomes sequenced to date, probably resulting from a drastic reduction suffered by this species after it split from its closest relatives. It may be useful to make a comparison with another species of the same genus, *Arabidopsis lyrata*. On the geological scale, the divergence from their last common ancestor is recent, at about 10 million years ago. During this time, the differences between the two species have become very considerable. The genome of *A. lyrata* is larger than that of *A. thaliana*: 32 700 genes are estimated in *A. lyrata* as against 27 000 in *A. thaliana*, but a quarter of the latter are not present in *A. lyrata*, half of the genome of which seems to have no equivalent in *A. thaliana*. We are a long way from the 1.2% gene differences estimated to separate humans from chimpanzees. The uniqueness of *A. thaliana* is also found in morphology, starting with the haploid number of chromosomes (see box on p. 12), which is just five, while it is eight in other *Arabidopsis* species and in Brassicaceae in general.

More seriously, given the role that *A. thaliana* has had and still has in developmental biology, it is poorly representative of plants in some aspects of genetic control of developmental processes. Genes involved in a combinatorial way in controlling the specification of the parts of the flower (sepals, petals, stamens and carpels) were first identified in *A. thaliana*. In its original formulation, the so-called ABC model of genetic specification of flower part identity envisaged that the flower organs where a gene function A is expressed

2 Cells and Development

The Cellular Basis of Development

Development Is Not Necessarily the Social Life of Cells

According to an incisive metaphor proposed in 1977 by the French developmental biologist Rosine Chandebois, morphogenesis, and development in general, is the product of the rules of cellular sociology. This image reaffirms the concept that the cell is the basic unit of development, and that understanding the properties and functions of cells is the main key to understanding the emergence of multicellular systems.

An echo of this statement resounds in the words of the British developmental biologist Lewis Wolpert, according to whom the complexity of development is an expression of the complexity of the cell. This sentence was not intended to mean that a single cell can undergo development, or that development in multicellulars is little more than development of unicellular organisms writ large. I will discuss development in unicellular organisms in a later section (p. 26), where I will present a different point of view from the traditional one. Here, let's continue discussing the less contentious issue of development in multicellular organisms. If it is true, as Wolpert adds, that what is going on within a cell is much more complicated than what is going on between cells in an embryo, what properties of a cell can make this a protagonist of development? Suggestions from specialist literature are very different and partly conflicting. Developmental genetics reveals that changes in gene expression are much more remarkable in particular moments of development and in the cells of particular regions of the body, a topic to which I will return in Chapter 6. But other studies focus on physical phenomena: Wolpert also remarked that morphogenesis is largely a problem of cell mechanics. This study should not be that difficult, as the types of force affecting cell behaviour reduce to a very short list, including tension, extension and changes of neighbours. In other words, it can be useful to look inside the cell, but also to observe its behaviour from outside.

In this section, the first of those dedicated to the aspects of development on a cellular scale, we will be content to consider cells as pieces of a mosaic: we will mainly consider their number, their size and some rules of mutual adjustment.

There seems to be no general rule to these things, as suggested by a comparison between the body cell composition of two groups of tiny marine animals: loriciferans and appendicularians. Loriciferans (whose first description, in 1983, was one of the most sensational zoological discoveries of recent decades) have an extremely complex structure, confined into just 300 micrometres (0.3 millimetres); their body, however, is made up of a very large number of cells, more than 10 000, a number of building blocks which seems to suggest a way to construct such a complex body. In contrast, the tiny body of the appendicularians (usually less than a millimetre) is made up of a very small number of cells: it takes less than a dozen huge cells to line their stomach.

The overall size of a plant or animal is affected by the environmental conditions in which it develops. However, the number of cells that make up its body, or its individual organs, is often irreversibly determined at an early stage in life. Let's consider gastrotrichs, tiny, little-known aquatic invertebrates, with adult body size between 70 and 1500 micrometres (0.07 and 1.5 millimetres) in length. This is the same range as in medium-large species of ciliates, but ciliates are unicellular whereas the body of a gastrotrich is made up of hundreds of cells. In freshwater gastrotrichs, all cells are already present at the end of embryonic development. Other animals in which the last cell divisions happen at an early developmental stage are briefly discussed in the next section.

In some cases, cell number and cell size seem to change in a coordinated and non-compensatory way. In the *chico* mutant of *Drosophila melanogaster*, the length of the adult reaches less than half the normal for this species. In this case, reduction in size is due both to the smaller average size of the cells and to their reduced number.

The model plant *Arabidopsis thaliana* tells a different story. If we leave aside the mutants characterized by unusual (generally reduced) size of leaves, flowers or flower parts, the size of the different organs of the plant is quite constant. This approximate constancy does not depend on a constancy in the number and size of the cells of which leaves, sepals or petals are made. Therefore, there seems to be a compensation mechanism: if the cells are smaller, their number will be higher; if larger, their number will be reduced.

A comparison between plants and animals, however, is not easy. In the stage of development that is called the embryo in plants, only two primary proliferative zones are identified, called the apical and radical meristems. The cells capable of going into mitosis remain confined to these two meristems, located at the two opposite poles of the seedling, while the cells that derive from them and eventually constitute stem and root do not divide further, but increase in size and differentiate. Throughout the life of the plant, the number of stem cells present in the two meristems remains practically constant. For the apical meristem, this number is around 1000 in the daisy species known as corn marigold (*Glebionis segetum*), but only 50 in *Arabidopsis thaliana*. The root meristem is much larger, including 125 000 to 250 000 cells. The continuous proliferation of these two meristems, however, is not sufficient to generate a complete plant, with leaves, flowers and, very often, lateral branches. The production of all these parts of the plant depends on additional proliferative centres, the lateral meristems. These, however, are not preformed in the embryo, but show up much later. In a centuries-old tree, new lateral meristems continue to form and to produce new leaves and new flowers.

These few examples from the development of plants and animals allow for a couple of general comments. First, if development is a matter of cell sociology, we must acknowledge that cell societies are extremely different from each other, both in the characteristics of the individual cells and in the nature of the rules governing their behaviour. Second, in some of these societies general laws dominate which impose compromises, as in the case of the relationship between number and size of cells in a leaf, while in others the degree of autonomy of individual cells is much greater, as we will see in the following sections.

Modulating Cell Division

When Cell Divisions Stop, Development May Not Also Stop

Important morphogenetic events are often preceded by a noticeable slow-down or even by large-scale suspension of mitotic activity. In *Drosophila*, for example, this happens at gastrulation, a process that requires extensive cell movements incompatible with cell division.

Increasing cell numbers, per se, is not a growth mechanism. Indeed, during mitosis it is probable that a cell will not easily be able to receive new supplies of nutrients, and its existing mass will be split between its two daughter cells. This is particularly true of the blastomeres deriving from the cleavage of a yolk-poor egg. As long as the number of blastomeres continues to increase, their average size continues to decrease. Conversely, increase in size at a time or in a body part in which mitosis does not occur is usual, both in plants and animals. When the lamina of a leaf of tobacco or *Arabidopsis thaliana* has grown to approximately a tenth of its final length, its cells cease to divide, starting from those closest to the distal tip, followed progressively by the others, up to the base. Cessation of mitosis is followed by a large increase in cell size and by cell differentiation.

In animals, stopping mitosis at a fairly early stage of development is often characteristic of some tissues or body parts. In humans, for example, striated muscle fibres and brain neurons lose all ability to divide after they have differentiated. In some zoological groups, mitoses cease at the end of embryonic development. This is observed in most mites and nematodes; in flatworms, mitoses stop at least in the cells of the epidermis and often in differentiated tissues generally. As we will see, however, in flatworms such as planarians there is a very abundant population of stem cells – that is, a reserve of undifferentiated cells ready to go into mitosis and to give rise to a progeny of differentiated cells. This guarantees the ability to regenerate lost parts (see p. 85).

A confirmation of the relationship that exists between the maintenance of mitotically active cells and the possibility of regenerating lost parts is offered by mites. As mentioned, in these tiny arachnids, mitosis almost always ceases at the end of embryonic development; however, in ticks, the largest of the mites, cells capable of dividing are also present in the post-embryonic stages.

Correlated with this, ticks can completely regenerate lost limbs, while this is not possible for other mites.

A corollary of the stopping of mitosis at the end of embryonic development is eutely – the constancy in the number of cells in the whole body of a small animal and in each of its organs specifically. As mentioned above (p. 16), eutely is known, for example, in small nematodes such as *Caenorhabditis elegans*. Even in animals where the adult is made of a number of cells that varies from individual to individual, such as arthropods, there can be constancy in the number of cells forming organs with very precise architecture, such as the ganglia in the nervous system.

Rather than being dependent on activation or suspension of mitotic activity, however, some aspects of early embryonic development appear to be dependent on the length of the cell cycle. This is often very rapid during cleavage. Consequently, the interphase between the end of a mitosis and the beginning of the next – the period in which replication of the DNA takes place, producing the copies that will be divided between the two daughter cells – becomes too short to allow gene expression. In some embryos, such as *Drosophila*, the rapidity with which divisions follow one another during cleavage is increased by the fact that the division of the nucleus is not followed by full separation into distinct cells; both nuclei remain immersed in a common cytoplasmic mass, and in a few minutes they are ready for the next division. In the appendicularians, the small marine invertebrates mentioned on p. 20, the duration of a cell cycle during cleavage can drop to as little as 4 to 5 minutes.

This frenzied sequence of nuclear divisions causes a delay to the first expression of the zygotic genome during early embryonic development: as a consequence, for a while, all morphogenetic events in the embryo are controlled by the expression products of the maternal genome (see p. 83). A longer interphase will eventually allow the expression of the zygotic genome, but for some genes the interphase must be quite large. In fact, even a very precise and efficient mechanism such as transcription may prove inadequate when challenged to transcribe a segment of DNA a million nucleotides long. This explains why a medium-length mitotic cycle may be insufficient for transcription. It is estimated that in *Drosophila* the transcription of the *Ultrabithorax* gene requires, at a temperature of 25 °C, a good 55 minutes, while in humans the transcription machine (which, at 37 °C, works twice as fast as its equivalent in the fruit fly) requires, at 25 °C, 11 hours to fully transcribe the very long *dystrophin* gene.

This is perhaps a good place to explain conventions in writing the names of genes. Gene names are written in italics, to distinguish them from the names of the proteins they code for: thus *dystrophin* for the gene, but dystrophin for the protein. The first letter is capitalized, except in the case of recessive mutants. This is the traditional usage, followed in this book, for animal, including human, genes. Conventions, however, are changing and are currently different for plant genes, where fully capitalized names are used (one example, YABBY, appears on p. 96).

Programmed Cell Death

Organisms Are Not Necessarily Built by Putting Cells upon Cells

In 1932, the 22-year-old Swedish palaeontologist Gunnar Säve-Söderbergh published the first description of *Ichthyostega*, a genus of fish-like vertebrates with four limbs apparently suitable for terrestrial locomotion rather than swimming. These fossils had been discovered the previous year in Greenland. Also found in Greenland were the remains of *Acanthostega*, described 20 years later, which again seemed to be transitional between an aquatic and a terrestrial condition in the evolutionary history of vertebrates. Both *Ichthyostega* and *Acanthostega* were found in rocks of the Middle Devonian epoch, dating back 365 million years. Other fossils, described in recent years, have helped in reconstructing this important evolutionary transition. A key step in this story was the precise reconstruction of their fingers and toes. In an article published in *Nature* in 1990, based on new material collected in 1987, palaeontologists Michael I. Coates and Jenny Clack revealed that the foreleg of *Acanthostega* ended with eight digits and the hindleg of *Ichthyostega* with seven. A few years earlier, the Russian palaeontologist Oleg A. Lebedev had described another fossil of this group, found in the province of Tula, south of Moscow. The forelegs of this animal, called *Tulerpeton* by its discoverer, had six fingers. Taken together, these fossils seemed to belong to a series that, starting with a higher and perhaps poorly fixed number of digits, would later stabilize as five, the number characteristic of today's terrestrial vertebrates (but often subject to further reduction).

However, since *Ichthyostega* and its relatives belonged to a phase of transition from aquatic to terrestrial life, it seems sensible to ask whether their toes were morphologically and functionally separated (like ours, or those of a lizard) or tied together by a membrane (as in the webbed feet of many water birds). Coates and Clack showed that in the hind limb of *Ichthyostega* there were four larger toes joined together by a membrane, plus three smaller toes, probably incapable of mutual displacement.

From fossils we can now move to embryos. In almost all terrestrial vertebrates, including humans, a growing hand or foot passes through a stage in which the fingers or toes, although by now outlined in their distinct skeletal axes, are joined together along the entire length: in other words, hand and foot are at this stage comparable to paddles. Separation of the digits is obtained through a process known as apoptosis, often also referred to as programmed cell death. Even those who, like me, find it inappropriate to use the word 'programme' in biology (p. 97) must acknowledge the regularity with which apoptosis occurs and the surgical precision of its effects. A hand with five separate fingers, for example, is the result of the loss by apoptosis of four rows of cells between one finger and the next in the original paddle-like embryonic hand. This does not happen in the foot of the duck, which is therefore webbed, and in all likelihood it had not happened yet in *Ichthyostega* and in the other vertebrates belonging to the transition phase from aquatic to terrestrial life. But even here, as is always appropriate in biology, it is necessary to refrain from generalizations. In salamanders, the forelegs end with (usually) four separate fingers, the hind limbs with (usually) five toes, but hand and foot do not go through a paddle-shaped phase: fingers and toes simply grow faster than the intermediate tissues.

In mammals we find important examples of apoptosis in the nervous system: in the cat, for example, more than 80% of the ganglion cells of the retina undergo programmed death. This phenomenon also plays a role in the annual process of shedding of deer antlers (p. 127).

In many invertebrates, apoptosis is involved in the destruction of larval structures at metamorphosis.

The most sensational case of apoptosis, however, is provided by colonial ascidians. Ascidians are marine invertebrates, most of which live as adults attached to rocks, shells and other substrata. Most ascidian species are solitary, but some are colonial. In the latter, a number of individuals, all deriving by asexual reproduction from a single founder, are physically conjoined and share a common vascular system. In *Botryllus schlosseri* (p. 77), each generation of these zooids is fated to die by programmed cell death, followed by phagocytosis (ingestion) of the dead cells by blood cells, while the subsequent generation of zooids reaches full development in a week.

Mechanisms different at the molecular level but comparable to programmed cell death in animals are also important in plants. The sap-conducting vessels are tubes made of large cylindrical cells aligned one on top of the other, the contents of which have been destroyed: perforations through the walls of contiguous cells allow the passage of sap. In some plants, localized destruction of cells leads to the formation of unusual leaves. In *Monstera* (a plant genus native to the tropical regions of the Americas, but familiar everywhere as an indoor plant), the leaf lamina is initially continuous but soon develops holes, due to selective histolysis (destruction of tissues) in some areas. A dissolution of cell walls is also the most important phase of the abscission process – the detachment of leaves, flowers and fruits from the plant.

The Development of Single-Cell Organisms

Development Is Not Restricted to Multicellular Organisms

During his long and productive life, the American biologist John Tyler Bonner has offered fundamental contributions to the understanding of the mechanisms underlying the transition from the single-cell to the multicell condition. This transition has occurred several times in evolutionary history, for example along the lineages that led to the emergence of animals and plants. A transition from a single cell to a multicellular organism occurs every day when the cells originating from an egg remain united together and take the form of an embryo. But the story can be different, as demonstrated by the cellular slime moulds (p. 36), a favourite subject of Bonner's.

There is a point, however, on which I allow myself to disagree with the opinion of this great scientist, that to study development is to study multicellularity. It is not difficult, in fact, to find striking examples of developmental processes in organisms formed by a single cell. I will give three examples here, the first from protozoans, the second from fungi (there are many single-celled ones, including bread and wine yeasts), and the third from bacteria. Further on (p. 58) we will see that development also extends to the production of gametes, the single-cell phase along the biological cycle of animals.

Let's start with protozoans, more precisely with the trypanosomes. This genus of tiny parasites is sadly infamous for the serious diseases that some of its species cause in humans (such as sleeping sickness and Chagas disease) or in domestic animals (such as nagana in cattle and horses). The biological cycle of these parasites does not take place entirely within the body of a vertebrate, but alternates between the latter and another host, which acts as a vector that carries the protozoan between two individuals of the vertebrate host species. Thus, the biological cycle of *Trypanosoma brucei*, the species responsible for sleeping sickness, includes stages in humans, where the parasite is usually found in the blood, and stages in the tsetse flies, hosted in their gut and salivary glands. Along its biological cycle, the trypanosome occurs in a number of different forms, which are easily distinguished (with a good microscope and simple staining techniques), especially on the basis of the presence, position and length of the flagellum (Figure 2.1). In trypanosomes the flagellum is characteristically folded back along the cell, which is elongated in shape, and thus forms a sort of membrane. The different developmental stages of trypanosomes are not separated by cell division. Thus, for example, the transition from the amastigote condition (without flagellum) to the epimastigote condition (with flagellum attached to the cell body by a short membrane) is a process of morphological differentiation, that is, a real developmental process. This process differs from the more usual developmental processes not so much because its theatre is a single-celled organism, but because it is reversible. But this also happens in some animals, as we will see in Chapter 4 (p. 64).

Another parasite, although not an obligate but an opportunistic one, is *Candida albicans*. This tiny fungus is very common on our skin or in our digestive tract: it is estimated that half of the entire healthy human population hosts it in the mouth or intestines. Like trypanosomes, candida also comes in different forms: the most frequent are a compact form, similar to the common yeast cells, and a filamentous one, a kind of hypha which immediately reveals the fungal nature of candida. The transition from yeast to hypha is usually induced by specific environmental conditions and is very rapid. Here too, there is no transition from one generation to the next: that is, no reproduction

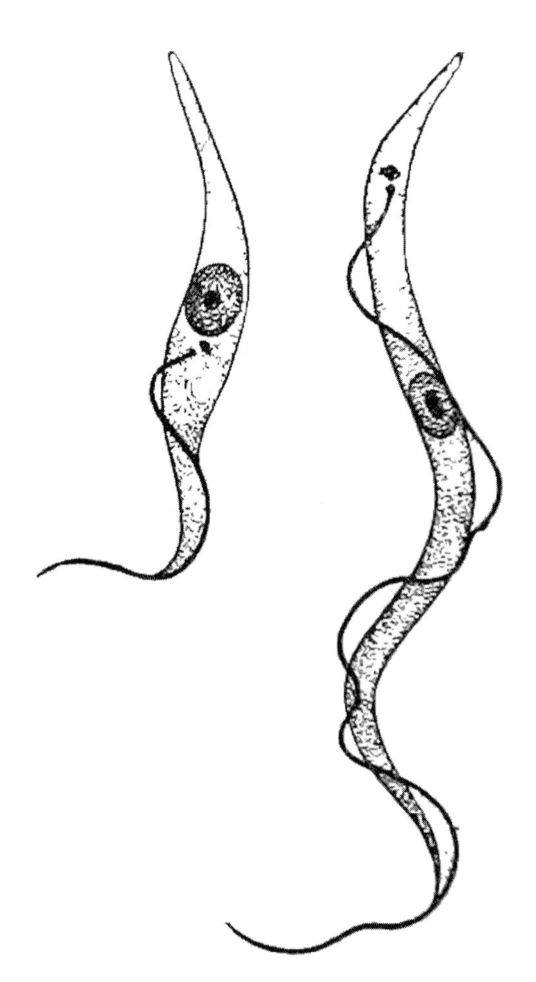

Figure 2.1 Two of the alternative cell shapes of the unicellular parasite *Trypanosoma brucei* (the agent of sleeping sickness): left, epimastigote; right, trypomastigote.

event between one stage and another, but an example of a developmental process exhibited by a unicellular organism.

Finally, let's move on to bacteria. Not all representatives of this vast world of unicellular organisms, whose cells are less complex than those of animals, plants and protozoans, have a simple form like the ball or stick shapes of the most popularly known bacteria; still more important, the shape is not necessarily constant in any given species. *Caulobacter vibrioides* is a free-living bacterium of freshwater environments. It has a characteristic shape, with a sort of tubular pedestal ending in an adhesive holdfast through which it is fixed to the substratum. When it divides, it gives rise to two very differently shaped cells: one is similar to the mother cell, the other is a swarmer equipped with a flagellum. Thanks to the latter, the swarmer moves for some time in the water, then fixes itself to the substratum, losing the flagellum and developing a peduncle with holdfast.

Acetabularia: Delayed Morphogenesis within One Cell

Morphogenesis Is Not Always Under Strict Control by the Nucleus

In our days, biology seems to be only a matter of genes and DNA. This is understandable, given the extraordinary development of molecular genetics in recent decades and the profound impact of its discoveries on medicine and agriculture. Unfortunately, however, many of the statements on everyone's lips are unjustified. We will dedicate the whole of Chapter 6 to the relationship between genes and development, but one particular aspect deserves to be addressed in the present chapter dedicated to the cell.

Whatever the actual role of genes in development, it seems necessary to admit that development depends on a continuous 'dialogue' between the chromosomes, which contain the so-called genetic information, and the other cellular components. Leaving out of this discussion bacteria, the only chromosome of which is not enclosed in a nuclear membrane, we might expect that there can be no development in the absence of a nucleus. This is confirmed by those cells that, in the course of differentiation, lose their nucleus, reaching a condition that no longer admits of any change except for the eventual destruction of the cell. Examples include the red blood cells of mammals and most of the neurons in the brain of some miniaturized wasps (the insect's whole body is a fifth of a millimetre long, and the average diameter of a brain cell in the adult is about 2 micrometres). But this is not always the case. We will see here two examples of single-celled organisms in which the relationship between nucleus and developmental processes is indirect, or even absent.

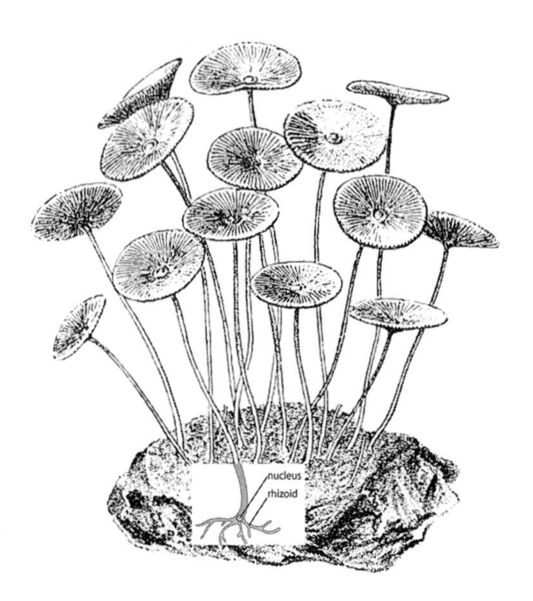

Figure 2.2 A group of individuals of the unicellular green alga *Acetabularia acetabulum*. The stem supporting the circular cap is up to 5 centimetres long. The insert shows the rhizoid through which the alga attaches to the substratum; the nucleus is located at about the transition from rhizoid to stem.

Let's start with *Acetabularia*, a genus of green algae of very unusual shape but quite common in the waters of temperate to tropical seas. These are algae of modest size, around 5 centimetres; but these dimensions are actually enormous, since these are single-celled organisms. When fully developed, *Acetabularia* have the shape of a tiny mushroom, with a cap supported by a long thin stem; at the pole opposite the cap there is a rhizoid, a kind of root thanks to which the alga is fixed to the substratum (Figure 2.2). At the time that it develops the cap, the *Acetabularia* enters the reproductive phase. The single nucleus, which is located in the rhizoid, undergoes a rapid series of divisions. The numerous tiny nuclei thus produced rise along the stem and colonize the cap, which divides into as many cysts from which a myriad gametes are eventually released into the water.

The nucleus, as said, is located in the rhizoid, but the most important morphogenetic processes take place during the life of *Acetabularia* (a year or more) at the opposite end of the stem, a few centimetres away. Even more remarkable is the fact that an *Acetabularia* deprived of its single nucleus not only survives for weeks, supported by the photosynthetic activity of its chloroplasts, but is also capable of forming a cap – in other words, of carrying out a complex developmental process. In some way, however, the nucleus is involved in these morphogenetic processes, even if its influence occurs at a great distance in space and time. During the whole life of the alga, a large number of messenger RNA (mRNA) molecules progressively accumulate in the upper part of the stem, but their translation into proteins, with consequent morphogenetic effects, will occur only under particular conditions, environmental or physiological.

Most of what we know about *Acetabularia* is due to the German biologist (of Danish origins) Joachim Hämmerling. His experiments on *Acetabularia* lasted more than 20 years. In 1943, Hämmerling demonstrated, for the first time in a eukaryote, that the genetic information is located in the nucleus.

In the next example, the role of the nucleus, and therefore of genes, in the production of complex structures is not indirect or delayed, as in *Acetabularia*, but literally absent. We move here to another group of unicellular organisms, the ciliates. A familiar textbook example is *Paramecium*, which consist of a large cell a fifth of a millimetre long. Many other ciliates are smaller, but some, like *Stentor*, can exceed 1 millimetre. Ciliates are decidedly unusual organisms, not only for their size or for the numerous or very numerous cilia that cover the cell body in whole or in part, but for the fact that they have at least two nuclei and for the unique character of their sexuality. The nuclei of a ciliate cell are of two kinds, at least one of each kind per cell – a micronucleus and a macronucleus. The first contains a 'standard' copy of the whole genome, the other is made up of many copies of short chromosomal segments. These have their equivalent in the micronucleus, but many chromosomal segments present in the micronucleus are not replicated in the macronucleus.

As for sexual phenomena in ciliates, these are not linked to reproduction; instead, reproduction consists here of a simple cell division, as in all cases of asexual reproduction of single cells. The sexual process (known as conjugation) is an exchange of nuclei that takes place between two individuals (the conjugants). These retain their structure practically untouched, including the cilia and the underlying scaffold of subcortical fibres. Their cytoplasmic connection is just a delicate bridge that breaks apart when the nuclear exchange has been accomplished. The two ex-conjugants, renewed in the genome but unchanged in the cytoplasm, are ready to begin their new life as independent cells.

In the late 1950s, the American researcher Vance Tartar performed an elegant series of micromanipulation experiments on the cell cortex of *Stentor,* in such a way as to obtain gross changes in the arrangement of the ciliary rows. At the following mitosis, the new ciliary pattern was faithfully inherited by the two daughter cells, and a morphologically stable clone of 'monsters' was eventually established. These modified stentors showed no tendency to correct the abnormal distribution of cilia. On the contrary, they became progenitors of a long series of generations of individuals in which the same anomaly was perpetuated. The behaviour of these cell lines was not entirely uniform and predictable, but the replication of the modified cortical structure along a number of generations is sufficient evidence of the substantial irrelevance of the genes in the gross morphological modification. Tartar's manipulation was not only limited to the outermost layer of the cytoplasm but, most importantly, it was purely mechanical. No intervention from the genes.

In *Stentor* and in ciliates in general, although perhaps not all of them, the morphogenesis of cortical structures does not depend on genes, but on the persistence of at least a fragment of the ciliary arrangement, which serves as a seed around which the construction (or reconstruction) of the whole pattern can take place. This occurs when the cell divides, passing on a portion of its cortical structures to each of the two daughter cells. Once the division is over, each of them will build the missing cortical structures to complete the pattern typical for the species. This process has some similarity to the replication of a DNA molecule, where each of the daughter molecules is made up of one of the filaments of the mother molecule, plus its newly made complement – except that, in the morphogenesis of the cortical structures of ciliates, DNA has very little to do.

A Lesson from Syncytia

The Building Blocks (Cells) Do Not Necessarily Precede the Building Processes

One of the reasons for the success of *Drosophila melanogaster* as a model organism is the speed with which this small insect completes its biological cycle. At 25 °C, it takes 10 to 12 days to go from egg to fly. Particularly speedy, as mentioned, is embryonic development, which takes just 1 day. Such rapid development is possible thanks to the syncytial condition of the fruit fly embryo up to the end of the 13th nuclear division. At this point, the embryo is not made up of distinct cells. Instead, there are about 6000 nuclei on the surface, surrounding a central mass of yolk within which other nuclei are dispersed, in modest number. Only 4 hours have passed since the initial division of the zygotic nucleus.

This concise description reveals three important aspects. One of these is the loss of synchrony between the dividing nuclei. After n divisions starting from the zygotic nucleus, there should be 2^n nuclei, but 2^{13} is 8992, a number significantly larger than 6000. It is clear that not all nuclei have kept pace with the fast dividing cycles undergone by the other nuclei. Latecomers are mainly the nuclei remaining inside the yolky mass; they are responsible for transcription and subsequent translation of genes encoding enzymes that help the mobilization of the yolk.

The second aspect to consider is that 13 successive divisions are accomplished in just 4 hours, equivalent to an average of less than 20 minutes per division: a speed more typical of a bacterial cell than of a 'normal' eukaryotic cell. This frenzied rhythm does not allow the DNA of these nuclei to become available for the synthesis of mRNA. This DNA is duplicated and the copies thus formed are immediately sorted between the two nuclei produced at the next division. As a consequence, in the embryo, at that moment, the DNA of its nuclei is not yet expressed; thus, the only available mRNA molecules and proteins are those accumulated in the egg before fertilization, that is, exclusively of maternal origin (p. 83).

The third remark is that no cells are seen in the *Drosophila* embryo until that point. The 6000 nuclei forming the surface film (the blastoderm) out of which the embryo will take form, gradually reabsorbing the central mass of yolk, are immersed in a common cytoplasm; therefore we describe this as a syncytial blastoderm.

Speaking of syncytia, in a developmental biology context it is important to maintain the useful distinction between (a) a situation like this, in which many nuclei are inside a common cytoplasmic mass because this has never been divided into cells after each nuclear division, and (b) the condition in which several nuclei are immersed in a common cytoplasmic mass following the fusion of cells that were initially distinct, with one nucleus each. This is the case with the fibres of our voluntary muscles.

In the *Drosophila* blastoderm, the syncytial condition is transient. After the 13th cycle, while nuclear divisions pause for a considerable time, the blastoderm becomes cellular: around each nucleus a cell takes form, delimited by the cell membrane. When this process is over, divisions resume, first in some groups of cells, then in others, in a precise sequence that allows the recognition of a number of mitotic domains, groups of contiguous cells in synchrony with each other but resuming divisions earlier, or later, than their neighbours.

Syncytial organization is also temporary in the embryos of other animals, with examples among fishes and sea stars. In many zoological groups, however, parts of the body are also syncytial in the adult. For example, the epidermis of the nematodes is syncytial: in *Caenorhabditis elegans*, one-third of the 959 somatic cells are fused to form 44 multinucleated syncytia. The peculiar tegument (neodermis) that covers the body of adult tapeworms and flukes is also syncytial. Muscles are often syncytial: this includes the skeletal muscles of vertebrates, as mentioned, but also those of insects.

In some cases, the syncytial organization seems to be induced by interactions with parasites or symbionts. Many insects, such as aphids, host within syncytial tissues symbiotic bacteria, from which they obtain nutrients. In freshwater bryozoans (little colonial polypoid animals) of the genus *Plumatella* attacked by a kind of very small parasitic fungus, the microsporidium *Schroedera airthreyi*, the peritoneum cells merge to form a syncytium. A similar effect is induced in their hosts (different marine invertebrates, such as molluscs, annelids and brittle stars) by the orthonectids, a small group of parasites that are probably close to annelids, but have evolved to be greatly simplified over their long practice of parasitic life.

The organization of the glass sponges (hexactinellids) is very special; these are sponges with a siliceous skeleton that are often described, somewhat improperly, as syncytial and therefore different from all the other sponges, called the cellular sponges. In fact, at the beginning of development, at least until the gastrula stage, the embryos of the siliceous sponges are formed by separate cells, like the embryos of the other sponges. Thin cytoplasmic bridges have been described as loosely connecting their cells, but it is only fair to mention that similar threads are observed in the embryos of squids and, as we will see shortly, in plants. Back to the siliceous sponges: as these progress in development, they become organized as a huge syncytial meshwork that houses in its holes a variety of specialized cells of different types; only thin connections remain between these and the syncytium.

Plants are considered cellular organisms, as the structural units of which they are made, well delimited by cellulose walls, have considerable functional autonomy; their cells, however, are permanently connected by plasmodesmata, cytoplasmic material that crosses the tiny pores with which the cell walls are pierced.

Again about plants, a question has been often raised, contrasting a cellular theory with an organismic theory of plant development: is the development of a plant the sum of processes mainly occurring at the level of the individual cells, or a global process that occurs in an organism made of cells? By experimentally hampering cell division, organs such as leaves can be obtained that are of normal shape but of smaller total size, even if made of larger cells. However, both theories represent extreme interpretations, while the truth, as often happens, seems to be in the middle. That is to say, plant cells have their own morphogenetic autonomy, but the architecture of the whole plant is determined to a certain extent also by signals that travel from one part of the body to another. This is true of animals and perhaps (p. 121) of fungi too.

3 Development as the History of the Individual

Many to One

Development Is Not Necessarily the Route from One Cell to Many

Slime moulds are organisms virtually unknown to non-professionals, which owe their name to their slimy look in one phase of their biological cycle and to the fact that they reproduce by means of spores, as do the true moulds. Together with the latter, slime moulds were long classified as fungi, but today, slime moulds and fungi are recognized as belonging to very distant evolutionary lineages. Moreover, there are two kinds of slime moulds, with no particular affinity between them. One kind is the syncytial slime moulds: some of them are patches a few centimetres across (as in the case of *Physarum*, of a lively orange colour, which we can find in the woods on rotting stumps) in which a very large number of nuclei coexist in a cytoplasmic mass not divided into cells. The other kind is the cellular slime moulds, on which I must spend a few paragraphs to describe the truly unusual way in which they go from a single-cell to a multicell condition.

I will focus on a species called *Dictyostelium discoideum*, which since the second half of the last century has been intensively studied by developmental biologists. Among these was John Tyler Bonner, as mentioned on p. 26.

The natural habitat of *Dictyostelium* is the soil, among decaying tree leaves, a humid environment where an abundant mass of organic material is available to a wide variety of organisms capable of feeding on it, such as many fungi and bacteria. Actually, the food chain in this environment is long and complex, and *Dictyostelium* has its place therein.

For most of its biological cycle, this cellular slime mould occurs as single cells. These are tiny (10 to 20 micrometre) amoeboid cells, which move very slowly on the surface of decaying leaves and other debris, ready to stop when they come across a bacterium, which they engulf and digest. The amoeboid cells reproduce by mitotic division, but the biological cycle of *Dictyostelium* does not reduce to perpetuation of the unicellular condition. Occasionally, there are episodes of sexuality, but here we do not need to deal with these in detail. Instead, let us see what happens when in their slow wanderings the *Dictyostelium* amoebae do not find any more bacteria to engulf. For such slow-moving organisms, the hope of finding more food before long is very meagre. But *Dictyostelium* still has one card to play. And it is here that the transition from the unicellular to the multicellular condition occurs.

These amoebae emit a chemical signal (in the form of cAMP molecules, cyclic adenosine monophosphate) that spreads in the surrounding environment; the molecule acts as a signal to which other amoebae of their species will respond. Therefore, even if their movements are very slow, it is likely that two amoebae will make contact sooner or later, thus forming a tiny two-cell cluster around which the chemical signal will be stronger than near a single amoeba. Over time, an increasing number of cells add to the initial cluster, until a critical mass is reached that leads to the emergence of a new behaviour. Without losing contact with one another, the grouped amoebae are now moving like a tiny slug, or a slimy mass. Soon, the small 'slug' stops moving and a mould-like shape appears. Of the many cells that make up the cluster, some remain in place, forming a disk that adheres stably to the substratum; the other cells climb on top of each other, rapidly forming a filament that rises to 1 to 2 millimetres in height. On top of this there gathers a mass of cells that surround themselves with a resistant coating: these are the spores, ready to be dispersed a few millimetres away, in new pastures hopefully rich in bacteria. Here the cycle will start again with a new generation of amoebae. Making filaments that carry a mass of spores is not an exclusive prerogative of the tiny cellular slime moulds, but a general characteristic of moulds – and this is precisely why *Dictyostelium* was once classified among the fungi.

What I have described is a developmental process in which a multicellular organism (the 'slug') takes shape by progressive aggregation of cells that until an instant before had led an autonomous existence. This story is very different from what happens, for example, during the embryonic development of animals. The egg (usually fertilized, but this aspect is not important for the present discussion) is a single cell that divides into two daughter cells, each of which divides in turn, and so on, rapidly giving rise to an increasing number of increasingly smaller cells that remain united together (with some interesting exceptions; p. 43). All the cells that form the body of an animal have a common progenitor, the egg. The same goes for the cells that make up the body of a seed-born plant.

The case of *Dictyostelium discoideum* and other cellular slime moulds is not truly unique. Five other protist groups have evolved a way of building multicellular organisms by aggregation; and examples are also found among the bacteria. However, these are isolated cases, which suggests that this way of building a multicellular organism is not very safe. In fact, there is a problem. Even if the distances over which they move are not such as to allow much mixing of populations, there is no guarantee that the amoebae that group together to form a 'slug' have identical genotypes. Any heterogeneity could be a problem. It is possible that in a heterogeneous aggregation there are cells more ready than others to reach the apical position on the filament, and therefore to turn into spores. Only these have any hope for the future, while the other cells, those that remain attached to the substratum, will end their existence there. If all cells in the 'slug' were descended from one progenitor, like those that derive from an egg, their genotype would be homogeneous; the same genome would therefore be present both in the cells forming the supporting stem and in the spores. Multicellularity by aggregation, instead, leaves room for competition that can translate into a trap for cells that have joined others more efficient at turning into spores.

The Biological Individual

A One-Cell Bottleneck Is Often the Beginning of a New Individual, but Not Always

As pointed out in the first pages of this book, in the popular notion (shared by many life sciences scholars), development is the sequence of changes that lead to the construction of an individual, starting from an initial cell – an egg, usually, or a spore, as in fungi.

In this book I am supporting a different notion of development; thus, I cannot fail to critically address the two fundamental tenets of the most common notion. One of these is that the living world is made of biological individuals; the other is that the life cycle of an organism in which the multicellular condition prevails must necessarily go through a single-cell phase.

The first of these tenets presents considerable difficulties, mainly due to an unjustified expectation. We expect, but we should not, to be able to extend without difficulty to all animals (even if distant by evolutionary kinship, body organization, forms of reproduction and developmental methods) all the notions that almost always apply unambiguously to humans and to animals such as mammals or birds. For example, not all animals have separate sexes, so the term hermaphrodite, originally used to designate a rare anomaly in our species, describes the normal sexual condition in leeches, earthworms and snails, as well as in vertebrates such as the sea bream and the groupers. The individual is precisely such a concept that does not apply easily to all animals, let alone plants. This issue – involving not only biology, but also the philosophy of biology – has given rise to a debate that has gained vigour in recent years.

The different notions of biological individual proposed thus far can be grouped around two different concepts. On the one side there are the notions of 'physiological individual', more intuitive and perhaps of larger practical interest, which recognize biological individuals based on objective criteria such as genetic uniqueness or physical separation from other individuals. On the other side there are more abstract notions, especially dear to philosophers of biology, who prefer to recognize evolutionary individuals, as, for example, entities that can compete with each other in a Darwinian logic. In the following lines I will focus only on some notions of physiological individuality.

Let us follow the lucid analysis of the French philosopher of biology Thomas Pradeu by asking three questions: what makes each biological individual unique? What are the boundaries of a biological individual? Finally, what are the moments or conditions that set the time limits for an individual's existence (or, better, its persistence)? It may seem easy to answer these questions. In the rest of this chapter and in further chapters, however, we will visit a gallery of animals and plants to which it is not possible to apply a notion of the individual based on the usual criteria.

Above all, however, we must deal with the very common instances in which the application of two different criteria of individuality leads to conflicting results. In many plants, for example, it is easy to take cuttings and thus produce physically separate but genetically indistinguishable individuals. To avoid the indiscriminate use of 'individual', the term *ramet* is used for each physically distinct plant, while *genet* is used for the set of ramets that share the same genotype.

To produce a plurality of ramets, human action is not necessary. In nature, many plants and a non-negligible number of animals habitually reproduce asexually. An example is the small hydra that takes shape from a bud (p. 76) bulging from the parent polyp.

If the ramets of the same genet demonstrate that physical separation is not necessarily a definitive criterion of individual uniqueness, other cases demonstrate that the absence of physical separation is not always a definitive guarantee of individuality. In this regard, there is no need to be fastidious. In what we can reasonably call individuals according to common sense (and also according to most of the criteria of biology), there is no guarantee that all cells have an identical genotype, especially if the individual is quite old. Indeed, mutations continue to accumulate inexorably. According to estimates published in 2012, every time human DNA is passed from one generation to the next it accumulates around 70 new mutations.

Things are different in asexually reproducing animals, with surprising consequences that have recently been described in the planarian *Dugesia subtentaculata*. This species, although also capable of reproducing sexually, usually multiplies asexually, by a process called fission. The new worm originates from a large number of cells from its parent, many of which are stem cells (neoblasts). As a consequence, the whole genetic diversity of the parent's neoblasts will be found in the new worm, which in turn will transmit it to its offspring. Between one fission event and the next, new mutations are sure to accumulate, thus increasing, even if only slightly, the genetic heterogeneity of the neoblast population and, therefore, of the whole set of cells in the next generation of worms. This steady increase in the genetic heterogeneity of the cells of animals that reproduce asexually is known as the Meselson effect (the name comes from the American geneticist Matthew Meselson). If an episode of sexual reproduction does not intervene to stop this accumulation, the genetic diversity of the neoblasts in a single planarian will increase to such a level that a criterion of individuality based on genetic uniqueness will no longer be applicable. According to the authors who documented this phenomenon in *D. subtentaculata*, individuality should not be ascribed to the worm, but to each of its neoblasts.

In less sensational cases, we can describe as a mosaic an individual in which two genetically distinct cell populations can be recognized, one of which is the result of a mutation that occurred during development. We must distinguish this from a chimera, which derives from the fusion of two or more individuals. We will meet with chimeras in a later section of this chapter.

Distinguishing between mosaicism and chimerism may prove difficult in the case of madrepores (stony corals), which are colonies of polyps joined together by a continuous sheet of living tissue. In a recent study of five different species of these corals, it was found that at least a quarter of the colonies (and almost half of the colonies in *Acropora sarmentosa*) are formed by polyps with different genotypes. These colonies are most often mosaics, but in each of the five species the researchers also found some chimeras. Those freshwater sponges that occasionally form by fusion of two or more larvae that metamorphosed close to each other on a submerged object are also chimeras.

In the case of sexual reproduction, the risks of producing an individual comprising a heterogeneous population of cells are limited to the effects of mutations that occur during individual development. The later these arrive, the lower the number of cells affected, also because – as mentioned on p. 23 – mitoses are frequent in the course of embryonic development, then tend to slow down and may stop altogether.

One or Many?

One Egg Does Not Necessarily Produce Only One Embryo

Undoubtedly, there is something special about twins. By this, I mean the so-called identical twins, those that derive from the same fertilized egg. Identical human twins inevitably attract special attention, not only from their parents and other family members, but from anyone who has the opportunity to interact with them. To contribute to the curiosity raised by the birth of identical twins there is their low frequency, estimated as about three births per thousand: this is the world average, but small communities are known in which the frequency of twins is somewhat higher. Never as high, however, as in the nine-banded armadillo, the best-known species of armadillos and the only one present in North America. In this species, females regularly generate identical quadruplets.

There are thus species in which this phenomenon – which takes the name of polyembryony – is the standard condition. This invites closer examination.

From a general point of view, one could discuss to what extent polyembryony is a matter of developmental biology rather than biology of reproduction. What happens, in fact? As in a normal embryonic process, the egg of a polyembryonic animal begins to divide into blastomeres, but soon these separate into two or more groups (four, as we have said, in the nine-banded armadillo), each of which continues to develop through the usual stages of embryonic (and post-embryonic) development. Thus, if the starting point is a single fertilized egg, while the end point is a multiplicity of individuals, it is difficult to deny that we are facing a reproductive process. Perhaps we may be reluctant to accept this interpretation, since in this case the production of two or more individuals does not involve sexual phenomena (production of gametes and fertilization) but has all the characteristics of asexual or vegetative reproduction. The latter is familiar to everybody as occurring in plants (in many species, it is sufficient to plant a cutting to obtain a new ramet) and in animals such as hydra, which usually reproduce by budding (p. 76). It is not easy to accept the idea that asexual reproduction can also occur in mammals and even in humans. However, it is better to interpret facts with an open mind, rather than restricting ourselves to preconceived interpretative schemes. The main problem is actually another. Polyembryony is one of those phenomena for which the divide between reproduction and development is not as clear-cut as in the most familiar cases. The issue becomes even more confused when a larva, rather than an early embryo, divides into two or more individuals. This happens in a few animals belonging to different zoological groups, from sea stars to tapeworms.

There are entire zoological groups in which we cannot expect cases of polyembryony, those in which the first division of the egg into two blastomeres produces an irreversible separation of fates. A complete embryo, smaller than normal, will never result from each of the two; at most, this will develop into an incomplete embryo destined for abortion. In Chapter 4 (p. 60) I will say more of this type of embryo.

The fact remains that any structure present in the embryo before it splits into two or more distinct units gets lost and must be built up again in each of the 'child embryos'. This problem is most serious when polyembryony starts quite late, as is the case for some insects, in which polyembryony takes on extraordinary proportions.

This happens in some tiny wasps, parasites of caterpillars. The first to notice this phenomenon was Paul Marchal, a French researcher who in 1898 described dissociation of the egg into multiple secondary embryos in the tiny hymenopteran *Encyrtus fuscicollis* (known today as *Ageniaspis fuscicollis*). However, the first accurate study was due to the Italian entomologist Filippo Silvestri, who in 1906 illustrated in extraordinary detail the polyembryony of a different species, *Litomastix truncatellus* (known today as *Copidosoma truncatellum*). Silvestri saw that from a single egg of this tiny parasitic wasp originated almost 2000 embryos: of these, more than three-quarters developed into normal larvae, each of which would transform into an adult, while the others developed into larvae without circulatory, respiratory, excretory and genital systems that failed to undergo metamorphosis.

In some animals, an embryo that begins to take shape undergoes fragmentation as in polyembryony, but in the end the fragments reunite into one embryo. In planarians of the genus *Dendrocoelum*, the eight cells deriving from the first three cycles of mitotic division disperse within the yolk, but eventually aggregate: this behaviour has been aptly named 'blastomere anarchy'. Somewhat similarly, in the small freshwater fishes of South America known as annual fishes (*Cynolebias*), two separate blastulae (more correctly, two blastoderms, as in fishes the stage equivalent to a blastula is an epithelial area on the egg surface) are often formed starting from one egg, each of which develops independently until advanced blastula stage. But when the two blastoderms begin to gastrulate, they unite again, and all their cells give rise to one embryo only.

Scaffolds

Developing Systems Are Not Merely Systems of Living Cells

The notion that all animals and plants are made of cells owes much to the work of two German scientists, the botanist Matthias Jacob Schleiden and the physician and physiologist Theodor Schwann. In a book published in 1839, Schwann proposed a 'cell theory' of the organization of living beings. The word cell, however, had been available in the life sciences since 1665, the year the English microscopist Robert Hooke published a richly illustrated volume titled *Micrographia*. Among the many objects observed with his microscope – a simple one, but powerful for its time – there was a thin slice of cork, which appeared to his eyes as a dense set of little boxes or, better, of small rooms: more precisely, he described them as little cells, exactly equivalent to the Latin *cellula*.

In fact, the little cells in the cork are just the empty walls, made of cellulose and suberin, of what had previously been a tree's living cells. Most of the trunk of any old tree is made of dead cells. This is true of the bark, which generally represents a modest fraction of the overall volume of the tree, but above all of the wood, which constitutes the entire central cylinder of the plant. The living cells are limited to a layer of modest thickness that lies between the woody core and the bark, and to this thin layer is entrusted the production of all the new cells that allow the tree's increase in diameter over the years. Of course, this simplified description does not include leaves and flowers, which are also made of cells. If we ignore these latter parts and consider, for example, a deciduous tree during the rest season, its living cells will perhaps be only 1% of the total.

In a sense, however, the dead cells of wood (rather less, perhaps, those of the bark, pieces of which are lost over time) still belong to the tree. They are not just the historical record of what in previous years had been complete, living cells. The woody core continues to offer a support to the plant's growth for as long as the tree survives. These dead parts are thus a scaffold for the development of the plant.

Scaffolds of different kinds are used by most living beings, perhaps by all of them. Like wood, many of these scaffolds are lifeless remains or lifeless productions of the individual — think of the shells of many eggs or the cuticles, of varying thickness and stiffness, of insects and crustaceans. But there are also living scaffolds. In viviparous animals, like most mammals including humans, the most important developmental scaffold is, of course, the mother's body. Scaffolds of a special type are the two partners – the fungus and the alga – of the lichen symbiosis, which reciprocally modulate their development (p. 50).

Chimeras

Chimeras Are Not Necessarily Unnatural

Harpy, Echidna, Sphinx and Medusa are four figures from Greek mythology whose names have been given to as many genera of animals; in their improbable company there are the dragon, the basilisk and the siren, whose legendary versions recur, often with very different traits, in the corpus of mythology of several peoples from different times. In these pages devoted to developmental biology, the most relevant among these monsters are two sisters, the daughters of Typhoon and Echidna, whose names are Hydra and Chimaera. In Chapter 5 (p. 76) we will consider the tiny polyps known today by the name of hydra; here we will deal instead with chimeras: not with the fish known as *Chimaera monstrosa*, also called the rabbit fish, but with the Chimaera of myth.

The Chimaera belongs to Greek, Roman and Etruscan mythology. This monster's aspect has been described in slightly different ways by authors such as Homer and Hesiod. The latter's description agrees best with the features of the splendid Etruscan bronze, cast perhaps 24 centuries ago, that was found in 1553 near of the Italian city of Arezzo and is now preserved in the National Archaeological Museum of nearby Florence. In this version, the head and the body of the Chimaera are those of a lion, but a goat's head protrudes from the back and the tail is replaced by a snake.

If we try to read the bizarre structure of the Chimaera in the light of modern biology, we run into a serious problem that puts the plausibility of its existence in question. The problem is the immune incompatibility between parts of the body of animals as diverse as lion, goat and snake. Over the past half century, we have witnessed progress in biomedical sciences that empowers us to mitigate and partially circumvent the problems that occur in transplants, but the case of the Chimaera seems hopeless.

The lion, goat and snake are vertebrates: that is, they belong to the lineage of organisms in which immune defences are the most advanced and effective, to the point that even grafts between individuals of the same species are usually rejected. Immune mechanisms are less effective, and less flexible, in other animals, but their presence, as far as we know, is nearly universal. Therefore, we should rule out developmental processes that would involve the fusion of genetically different individuals. However, some facts force us to change our mind.

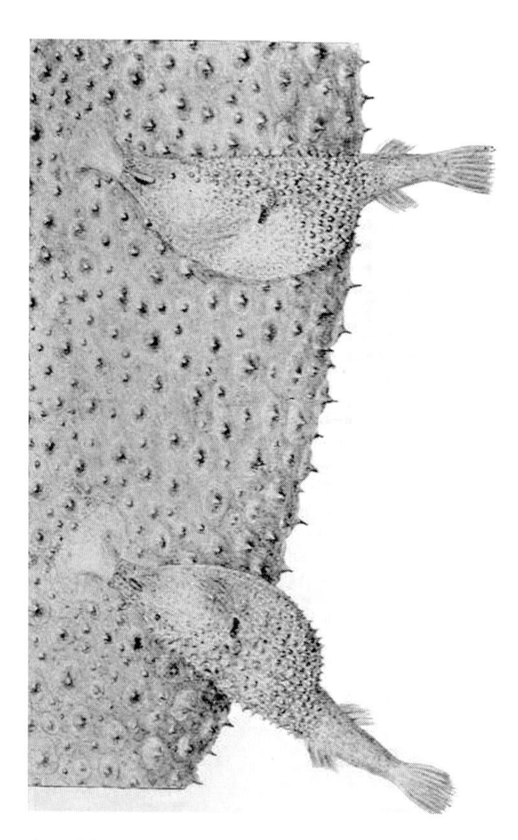

Figure 3.1 Two dwarf males of the sea devil *Ceratias holboelli*, attached to the skin of a female a dozen times longer than the males. From an illustration in a 1922 paper by the Icelandic naturalist Bjarni Sæmundsson, who wrongly interpreted the small fishes as offspring of the female. The correct interpretation was first provided 3 years later by the British ichthyologist Tate Regan (see text).

The first story is particularly striking because it has vertebrates as its protagonists. It begins with a 1925 article by C. Tate Regan, Keeper of Zoology in what was then called the British Museum (Natural History) – today, more simply, the Natural History Museum. Regan said that he had found some dwarf male fishes attached as parasites to the bodies of females of their species (Figure 3.1). The discovery involved two species (*Ceratias holboelli* and

Photocorynus spiniceps) but, as Regan rightly suspected, this behaviour is shared by all Ceratioidei, monstrous-looking abyssal fishes more commonly known as sea devils, or deep-sea anglerfishes. In these fishes, the differences between the two sexes are extreme, primarily in size. In *Ceratias holboelli*, females are up to 60 times the length of the conspecific males. In some species, the male behaves as an external parasite of his partner, but in others the epidermis and dermis of the two partners merge and the circulatory system of the male joins that of the female. In this way, a real hermaphrodite chimera is created. No immune reaction? Apparently none. We do not have any specific information on these abyssal fishes, but a recent discovery about one of their relatives living in shallower waters, the monkfish (or anglerfish) *Lophius piscatorius*, throws light on this story. This species (in which males lead a normal, free life) lacks all the genes necessary for the construction of the major histocompatibility complex (MHC) class II pathway. It is therefore probable that the loss of this fundamental immune defence tool preceded, in the history of this fish group, the evolution of the unusual sexual behaviour of one of its more specialized branches, the deep-sea anglerfishes.

Pairs of inextricably bound individuals that integrate their organs and tissues are also found in two kinds of parasitic flatworm. We do not know anything about them from the point of view of the immune system, but in this group of invertebrates there is no reason to believe that they have complex and sophisticated immune mechanisms like those of most vertebrates. The most interesting aspect of these worms is the fact that each of the two partners is hermaphrodite, like the vast majority of flatworms. Being a hermaphrodite, however, does not always guarantee reproductive autonomy (although this happens in the common *Taenia solium*, the pork tapeworm, known in some languages as the solitary worm). In flatworms of the genus *Diplozoon*, external parasites of freshwater fishes, both partners have perfectly functioning male and female apparatuses and, once two individuals have united, each will fertilize the eggs produced by the other. In another of these worms, *Wedlia bipartita*, the larger partner produces only eggs, while the male reproductive system degenerates; the opposite occurs in the smaller partner.

Even more unusual is the way a male/female chimera is formed in the rhizocephalans. These are crustaceans parasitic on other crustaceans. Their shape is so odd that it is extremely difficult to guess that they are crustaceans

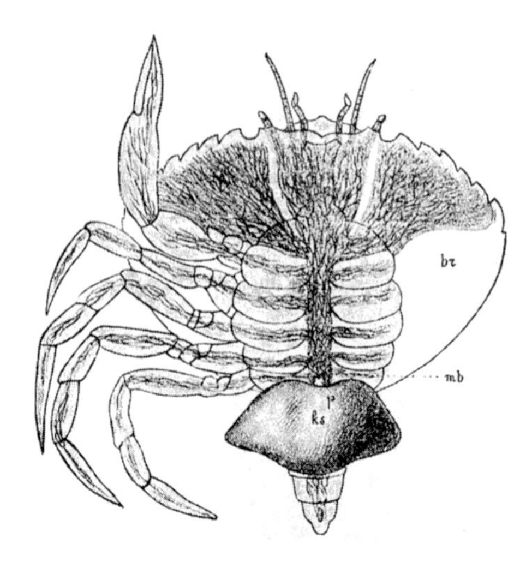

Figure 3.2 Ventral view of a crab parasitized by *Sacculina carcini*. The large sac is the 'externa' of the parasite, where male and female gametes are produced; it is attached to an extensively branching system of 'roots' depicted here as seen through the crab's cuticle.

or even arthropods (at least in their reproductive stage; their larvae are not that different from those of many other crustaceans).

The first description of one of these bizarre parasites dates back to 1787, when Filippo Cavolini, an Italian naturalist from Naples, described it and rightly identified it as a kind of crustacean, but did not provide a name for it. The animal was described again in 1836 by the British military surgeon and naturalist John Vaughan Thompson, who gave it its current name of *Sacculina carcini* – literally, the crab's small sac.

A crab parasitized by a *Sacculina* shows obvious signs of weakness, and its secondary sexual characteristics are often faded: this is a sign of so-called parasitic castration. The presence of the parasite, however, is manifested above all by the large bag that protrudes between the appendages of the crab, in the space enclosed by the folded pleon (the "abdomen"), which in the female crab is used as an incubation chamber. The protruding bag (the 'externa' of the parasite; Figure 3.2) is attached to the victim through a sort of extensively branched root that deepens into the body of the crab. The externa is provided with a tiny opening from which the very small larvae of the parasite can escape. A *Sacculina* produces both eggs and sperm, but this parasite, like the sea devils, is actually a hermaphrodite chimera.

As said, in the larval state *Sacculina* is not much different from many other crustaceans. Its post-embryonic development begins with a stage of nauplius, with a very short body and three pairs of appendages only. The nauplius turns into a second stage, with a greater number of appendages, called a cypris. This is the infectious stage, at which the two sexes are still distinct, although male and female cypris look very similar. The interesting part of the biological cycle begins when a female cypris attaches, through one of its antennae, to a robust hair (a 'seta') protruding from the cuticle of a crab. The cypris progressively loses abdomen and thorax, eventually reducing to a mass of cells that deserves to be called embryonic, which passes into the body of the crab through the thin tube formed by the antenna of the cypris. Once the victim's internal organs are reached, in the vicinity of the intestine, the parasite's cells proliferate, forming a sort of tumour that is gradually pushed backwards and is eventually extruded to form the externa.

Up to this point, the *Sacculina* that is developing at the expense of the crab is a female: but how can sperm also be produced in the externa? This becomes possible because a male cypris is able to recognize a virgin female externa and to attach itself to it. Within 20 minutes, from the male cypris there emerges a trichogon, a tiny amoeboid mass without appendages, muscles or sense organs, or even a nervous system. Two hours later, the trichogon joins the female cells already present in the externa; and a few days later, the production of sperm cells begins. The male part of the biological cycle of *Sacculina* was discovered only in the 1980s, mainly by the Danish zoologist Jens T. Høeg: 100 years had passed since the French zoologist Yves Delage discovered the female part.

From the point of view of developmental biology, the cases summarized in this section are the most sensational, but before moving on to a different topic it is worth taking a look at plants. Here chimeras are literally everywhere.

In flowering plants, up to three generations remain in anatomical continuity in the transition from flower to fruit. The haploid female spore or megaspore remains enclosed in the place where it was produced. From its proliferation, a tiny haploid gametophyte is formed, typically of seven cells one of which has two nuclei: this is the ovule+embryo sac complex, which also remains enclosed in the ovary. If the ovule is fertilized, the diploid zygote will produce the seedling (also diploid), which with its envelopes and the associated triploid endosperm (a nutritional tissue; p. 59) forms the seed.

Finally, any moss plant is a chimera when, on the leafy plantlet, a thin filament develops, ending with a capsule in which the spores differentiate. The cells forming the leafy plantlet are haploid, as is the spore from which it originated, whereas the filament and the capsule are part of a distinct generation, formed by diploid cells, which originates from the fertilization by a male gamete of a female gamete produced by the haploid plantlet. The female gamete remains solidly fixed to the haploid plantlet (the gametophyte) from which it has differentiated, as does the diploid plantlet (the sporophyte) that arises from it, following fertilization.

Lichens and Plant Galls

Two Organisms May Not Have Two Developmental Histories; Sometimes It Can Be Three

An organism is unlikely to develop in a sterile environment that protects it from any influence by organisms with a genome other than its own. Influences are not necessarily adaptive and rarely so specific that they leave a characteristic mark in morphogenesis. However, there are two important cases in which a close association between two very different multicellular organisms leads to the production of forms with their own constant characteristics, sometimes very complex ones.

The first example is lichens. Lichens are undoubtedly odd things; it has not been easy to place them in the classification of living beings. For a long time, they were considered a particular group of plants, but this kingdom was intended in a broader sense than it is today: in particular, it also included fungi. An affinity of lichens with fungi was long suspected, but their dual nature was discovered only in the 1860s: each lichen is, in fact, a close association between a fungus and an alga. This discovery is mainly due to the Swiss botanist Simon Schwendener and the German surgeon, botanist and phytopathologist Heinrich Anton de Bary; in 1879, the latter introduced the new term symbiosis.

It was soon clear that the algae involved in lichen symbiosis belong to a few genera only, of which additional species that do not live in association with fungi are also known; in contrast, the fungi involved in lichenic symbiosis are very numerous and almost all very different from non-lichenizing species. Moreover, the shapes of lichens, although very different from those of fungi not associated with algae, are mainly determined by the fungal component rather than by the algal one. For these reasons, systematists have decided to classify lichens based on the fungal partner, and today they are distributed in a number of groups within the classification of the Ascomycetes, with the exception of a few dozen species classified with the Basidiomycetes. Some of these groups include lichenizing fungi only, others also include related non-lichenizing fungi, a close relationship with the former being often recognized based on DNA analysis.

We might search in vain for a chapter on lichens in developmental biology books, but they would certainly deserve one.

To briefly illustrate the biology of lichen development, it is necessary to start from their reproductive biology. Lichens can reproduce both asexually and sexually. In the first case, the symbiosis is maintained, because the reproductive units (simple fragments of the parent lichen or, more often, specialized entities called soredia and isidia) include both fungal hyphae and algal cells. But a lichen's sexual reproduction is actually the sexual reproduction of the fungus. The process, therefore, presupposes the breaking of the symbiosis with the alga, which will be sealed again in the next generation.

In this cycle of dissolution and reconstitution of the lichen symbiosis, the continuity we may require as a condition for being a biological individual is lost. Other properties, such as the functional integration of the parts, may be saved. This is true, however, only in the most typical situations, from which many lichens depart considerably, either occasionally or regularly. For example, the degree of integration between the two partners can be very different, not only at different times in their integrated development, but also in different parts of what from a morphological point of view we describe as an individual thallus. For example, a thallus of *Aspicilia californica* is composed of branched axes, with central medulla and bark formed by fungal hyphae, while the intermediate layer is made up of algal cells. The tips of the branches, however, are formed exclusively of fungal tissue.

Sometimes, hyphae that have developed from the fungal spore come into contact here and there, on the substratum, with groups of cells of an alga with which they can enter into symbiosis; consequently, at least for a while, the system includes both lichenized and exclusively fungal parts. Finally, in one and the same thallus, different algal and especially different fungal genotypes can coexist. Describing these cases in terms of biological individuals is particularly difficult and in any case subjective.

One of the reasons to dwell a little on lichens is their obvious evolutionary success. This is evidenced not only by their diversity (no less than 17 000 species of lichenizing fungi have been described so far, and it is estimated that those still to be described are much more numerous), but also by the antiquity of the lichen symbiosis. Filamentous hyphae closely associated with algae or cyanobacteria have been found in rocks of the Doushantuo Formation (between 551 and 635 million years old) in Weng'an, in the south of China.

On a geological scale, the production of galls caused by plant/insect interactions is much more recent: the oldest fossil records are around 100 million years old. But galls, like lichens, are morphologically very diverse. Quite often, their shape is so characteristic and specific as to allow identification, like the specimens of a biological species. Particularly specific (and often complex) is the shape of the galls induced by hymenopterans of the cynipid family. Many species of these little wasps produce galls on oaks. A single oak species can host up to 40 different gall types, produced by punctures made by as many species of cynipids.

A comparison between galls and lichens is interesting. In both cases, a very precise and sometimes complex form (the thallus of the lichen, or the gall) is produced, one generation after another, by a system consisting of two distinct organisms, each with its own genome, which continuously dissociate and re-associate. In other words, there is no 'gall genome', just as there is no 'lichen genome'. Nevertheless, it seems legitimate to include both lichens and galls in the list of biological systems in which development occurs. This choice remains valid even if we demand (though perhaps we should not, as discussed in Chapter 5, p. 93) that development must be adaptive. In the case of lichens, it is enough to remember that they are one of the best examples of mutualistic symbiosis (in which, that is, both partners benefit from the association). In the case of galls, the advantages for the two partners are different and partly opposite. The advantage for the insect is more obvious, as the gall offers protection and nourishment during the insect's embryonic and larval development. As for the plant, it has been suggested that the production of a gall may help to restrict the insect's attack to stem or leaf tissues.

4 Revisiting the Embryo

Developmental Inertia

Development Does Not Always Have a Recognizable Starting Point

Susan Oyama is a psychologist and philosopher of science whose books have found considerable resonance in the debate on central topics of biology such as development, evolution and inheritance. Her name is associated with developmental systems theory (DST), according to which a distinction between the contributions of genes, environment and epigenetic factors to developmental processes is necessarily arbitrary. In agreement with current usage, rather than with the historical one (p. 90), by 'epigenetic' I mean here any heritable change in the phenotype (the term introduced in 1911 by the Danish botanist Wilhelm Johannsen for the observable traits of an organism, as distinct from its genotype) that does not involve permanent alterations in the protein-coding sequence of nucleotides in the DNA molecule.

As expected, in this context, Oyama has taken a position on the role of genes in controlling developmental processes. In her book *The Ontogeny of Information*, she reacts to the popular gene-centric view of development by remarking that a gene can be considered to initiate a sequence of events only if our investigation starts at that point. This statement downsizes the role of genes in development and, above all, denies the existence of those 'master control genes' (discussed in Chapter 6, p. 100) which according to some authors would have under their control the morphogenesis of entire parts of the body, like an eye or a heart.

Oyama's remark, however, carries another message, with which I open this chapter: we can choose to describe a sequence of developmental processes starting from any time convenient to our investigation, but we must not forget that this choice is arbitrary, because development is a continuous process, and has been ever since there was life on Earth. In many respects, it is useful to restrict attention to what happens over the course of one generation, but we cannot accept that everything starts from scratch at fertilization.

According to Oyama, all developmental stages are more or less arbitrary temporal slices through the life cycle. The phenotype exhibited by an organism at a given stage depends on multiple causes and agencies, including gene transcription at previous stages, mechanical influences and past activities, and it has certain prospects for further change. This applies not only to the times we usually regard as intermediate, such as gastrulation or metamorphosis or the achievement of sexual maturity, but to any stage, including the one we traditionally regard as initial: the egg (any egg) is full of history, and only a part of that history is written in the genes.

The long history whose first stages should be sought in the most remote manifestations of life and which continues in each living being is, of course, evolutionary history, but its backbone is the continuity of development through the generations. During this very long history, there are both the short-term cyclically repetitive changes we describe as development, and the long-term changes we call evolution.

This integrated vision of development and evolution opens a promising and probably unexpected window. Let's start from evolutionary biology, where theoretical models have been developed much more extensively than in developmental biology, perhaps compensating for the lesser accessibility of the phenomena to be studied. This largely depends on the different time scales: developmental segments of plant and animal life are there for all to see, but evolutionary change is not that readable, even where it would be easiest, as in the emergence of new virulent bacterial strains under the selective pressure of antibiotics.

A milestone in evolutionary biology is the Hardy–Weinberg principle, according to which, in a population, the frequencies of the different alternative forms (the alleles) of a gene remain unchanged from generation to generation in the absence of evolutionary influences such as mutation, natural selection, population bottlenecks, migrations and others. We therefore speak of Hardy–Weinberg equilibrium. This does not represent a description of the conditions observable in real populations, which deviate – some more, some less – from equilibrium. However, it is possible to make a substantially complete list of the conditions under which an ideal population would be in Hardy–Weinberg equilibrium, such as random interbreeding, no mutation and no selection. By knowing these conditions, it is possible to investigate the reasons for the departure of a real population from that ideal behaviour. Technically, we can say that the Hardy–Weinberg equilibrium is the inertial model for evolutionary biology.

Inertial models are better established in sciences other than biology. Newtonian mechanics, for example, rests on an inertia principle: every body, as much as in it lies, endeavours to preserve its present state, whether it be of rest or of moving uniformly forward in a straight line. Of course, a body in inertial conditions is not very interesting, but the principle is important as it allows us to afford the study of the movement of actual bodies. Any deviation from the inertial behaviour is the result of constraints or forces applied to it. The similarity to the above discussion of the Hardy–Weinberg principle is obvious.

That said, it is time to go back to developmental biology. Is it possible to define here a principle comparable to Hardy–Weinberg's? Or, to use the language of physics, is it possible to identify a hypothetical inertial behaviour in development?

I think that this is not only possible but also useful if we want to get closer to understanding the role that the most diverse agents can have in development, from the expression of genes to environmental influences.

According to the American philosopher James Griesemer, such a principle should eventually be defined in adequately general terms, without explicit reference to organisms with cellular organization. However, from an operational rather than a philosophical point of view, we can probably be content with defining the inertial dynamics of development in terms of uniform growth in space and time – that is, as an indeterminate iteration of cell proliferation.

In many multicellular organisms, the short phases of intense cell proliferation observed in the cleavage phase of embryonic development are followed by a prolonged arrest of mitotic activity. As mentioned in Chapter 2 (p. 23), this arrest is sometimes generalized and irreversible, as in many mites and nematodes, but it can also be selective, as in earthworms, where mitosis is stopped early on only in the epidermis.

If the inertial condition is the indeterminate proliferation of a cell line that always remains the same, examples close to this model are found in the behaviour of stem and cancer cells. Interestingly, one of these examples comes from cells that we consider part of a coherent, adaptive developmental system, the other from cell lines we consider disruptive. This is a good sign of the inertial model's ability to serve as a starting point for the study of developmental processes regardless of their adaptive value. I am reminded of an observation of Charles Darwin, concise, like all the notes in his *Notebooks*: "most monstrous form has tendency to propagate ... Even a deformity may be looked at as the best attempt of nature under certain very unfavourable conditions."

In nature, however, all processes deviate more or less conspicuously from this hypothetical inertial behaviour. Even stopping mitosis is a deviation, with its specific causes, and a single cell, in my understanding (p. 26), can also undergo development without proliferating. But that is just one aspect. Sooner or later, signs of differentiation appear among the blastomeres, which concern not only the size, shape and mutual position of these cells, but also the set of genes transcribed therein. Cell proliferation no longer proceeds indifferently, limited only by the availability of material and energy resources. This limitation can be considered to be a cause of deviation from the inertial behaviour; but there is much more – the whole constellation of influences that we are used to separating into 'genetic' and 'environmental', perhaps including in the latter the effects of contiguous cells or cells further away that still influence behaviour through the diffusion of molecules. The subdivision between environmental and genetic determinants may be convenient, but it is a gross simplification of reality. This seems to be a reasonable conclusion, even if we may prefer not to embrace all the implications of Susan Oyama's DST.

From the Zygote, but Not in the Embryo

Some of the Cells Deriving from the Zygote May Not Contribute to the Embryo

Let us focus once again on the first stages of development of a multicellular organism. If this is an animal, the process will usually start with a fertilized egg, or zygote; from this, through a series of mitoses, an increasing number of cells originate, which make up the embryo. In many cases, the embryo can be considered an autonomous organism, whose exchanges with the outside do not go beyond those required by cellular respiration. For a while, the food requirements of the new individual are met by the reserves stored in the egg – the yolk accumulated there during oogenesis (the production of eggs in the mother's body). Often these reserves are modest, so development is possible only on condition that the animal soon gets nourishment from the external environment. This is the case with many marine invertebrates, in which the embryo develops into a tiny larva with mouth and digestive cavity. In other cases, yolk reserves are more abundant and provide for prolonged autonomy. This is the case, for example, with birds, where the nutritional resources available to the embryo are sufficient to support development into a chick.

However, things are not always that simple. In most flatworms, such as tapeworms, flukes and planarians, a good amount of yolk is available, but not in the egg. The ovaries of these animals (which are almost all hermaphrodites, but here we are only interested in the female reproductive system) produce small eggs, without yolk; yolk is instead accumulated in cells produced by separate organs, the vitellaria. Eggs and yolk cells will be associated (but not merged together) before the egg is covered by a shell. The animal therefore produces compound eggs.

The opposite case is more frequent, in which the yolk initially found in the egg is not distributed among all the cells that derive from the embryonic divisions but remains confined in a few larger cells. These do not participate in the formation of the embryo but are like a store from which yolk is gradually resorbed. In *Drosophila*, as in many other insects, yolk is located in the centre of the egg; the embryo takes shape on the surface of a yolky mass that does not divide into cells, but remains syncytial.

This mass, however, is not the only derivative of the insect egg that does not take part in the formation of the embryo. Two extra-embryonic membranes, amnion and serosa, also originate from the zygote. These membranes are more developed in insects with incomplete metamorphosis, such as grasshoppers and cicadas; at least one of these membranes, however, is also recognizable in insects with complete metamorphosis, such as bees and flies.

The production of extra-embryonic membranes and other accessory structures accompanied the successful evolution of two large zoological groups out of water. In addition to insects, this innovation also appeared in the amniotes, the evolutionary line of vertebrates that includes mammals and reptiles (among the latter are the birds which, as is now widely known, are a kind of surviving dinosaur): that is, both viviparous (live-bearing) and oviparous (egg-laying) animals. The embryonic annexes of the amniotes are the amnion, the allantois and the yolk sac. The use of the same term (amnion) for insects and vertebrates, of course, should not be taken as an indication of kinship, but simply as the name of structures that have a similar function – in this case, to enclose a cavity full of liquid that surrounds the embryo. In addition to the embryo and these embryonic annexes, a mass of cells called the trophoblast also develops from the zygote of the mammals, which makes contact with the uterine wall and, together with maternal cells, contributes to the production of the placenta.

As in animals, in plants too it is not true that the embryo corresponds to the whole set of cells derived from the zygote, and only to those. For brevity, here I will continue to use the term embryo for plants, irrespective of the critical remark in Chapter 1 (p. 14).

In flowering plants, the first division of the zygote gives rise to two different cells: the seedling will form from the smaller of these, while the larger gives rise to the suspensor, a temporary structure that allows the passage of nutrients to the embryo. We need to devote a few words to the origin and location of these nutrients.

To clarify ideas, we need to say a couple of words on the reproductive biology of flowering plants. In their biological cycle, as typically in organisms with sexual reproduction, there is alternation between a haploid and a diploid phase, but the former is not as ephemeral as in animals (although it is much reduced compared with the haploid phase in the life cycle of mosses and ferns). Haploid cells produced by meiosis do not behave like gametes, but undergo a small number of cell divisions (or, at least, of nuclear divisions) that give rise to tiny multicellular organisms called gametophytes. Only one of the cells of each gametophyte behaves as a gamete. The male gamete is one of the three cells of a pollen grain, whereas the female gamete is one of the seven cells of what is called the egg+embryo sac complex. To complicate things, two other cells intervene, one male and one female, which join together in what we may call a second fertilization. On the female side, the cell involved is the so-called central cell, which contains two haploid nuclei. As a result, a sort of triploid zygote forms from its union with a male haploid cell. It will give rise to a triploid tissue called the endosperm. This will feed the embryo through the suspensor.

Chromosome numbers aside, the origin and fate of the endosperm of flowering plants are somewhat reminiscent of the yolky cells of flatworms, but also of the food resource exploited by adelphophagous animals. In these viviparous animals, an individual feeds at the expense of other embryos present in the mother's genital tract – hence the name of the phenomenon, which means 'eating a brother'. Known examples are not numerous, but this behaviour is found in species belonging to disparate zoological groups, that is, snails, insects, crustaceans, echinoderms and also vertebrates – the fire salamander (*Salamandra salamandra*) and the bull shark (*Carcharias taurus*).

Order Emerging

Development Does Not Necessarily Produce an Increasing Division of Labour

A traditional issue in comparative embryology is the contrast between mosaic development and regulatory development. This distinction originates from the contrasting results obtained in the last two decades of the nineteenth and the beginning of the twentieth century by two German researchers, Wilhelm Roux and Hans Driesch. In an 1888 experiment, Roux used a red-hot needle to kill one of the two blastomeres deriving from the first division of a frog egg and saw that a half-embryo developed from the surviving blastomere: the parts that would have derived from the destroyed blastomere were missing. This experiment, and others of the same type, performed at the next stage of development (four blastomeres), convinced Roux that the successive divisions the embryo undergoes partition it into an increasing number of cells in each of which remain confined the nuclear factors corresponding to a specific part of the body. In other words, the embryo appears to be a mosaic of building blocks (the blastomeres), each with an irrevocably assigned morphogenetic fate.

Very different results were obtained by Driesch, who conducted similar experiments on the sea urchin (Figure 4.1). These new experiments were more precise, for two reasons. First, the sea urchin egg and the resulting

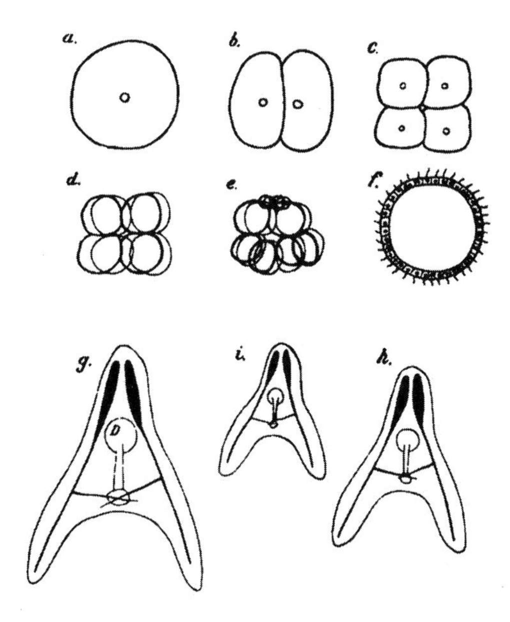

Figure 4.1 Hans Driesch's experiments on sea urchin embryos, in a series of very rough drawings published by the German embryologist in 1905. (a–f), Embryonic development from the zygote (a), through two-, four- and eight-blastomere stages (b–d), morula (e) and blastula (f). The embryonic development ends with the production of a pluteus larva (g). By dissociating embryos in the two- or four-cell stadium, Driesch obtained from each blastomere smaller but morphologically regular larvae (h, from one blastomere of an embryo as in b; i, from one of the four cells of an embryo as in c).

embryo are transparent, making observations under the microscope easier and more precise; second, in the sea urchin it is possible to separate the blastomeres mechanically (without killing them) as long as they are quite large and few in number – that is, after one of the first divisions. In this way, Driesch was able to obtain isolated blastomeres not attached to a dead cell, as in the experiments performed by Roux. This was a major improvement, because the proximity of the dead cell might have influenced the further development of the preserved blastomere. Whatever the cause, by observing the development of a sea urchin blastomere separated from its twin at the two-cell stage, Driesch did not obtain a half-embryo, but an embryo (and then a larva) that was smaller than normal but complete. A similar result, except for the even smaller size of the resulting embryo or larva, was obtained by eliminating three of the four blastomeres resulting from the second division. Unlike what Roux had observed in the frog, in the embryo of the sea urchin Driesch did not find an early division of potential between the individual blastomeres. At least in the very early stages (two to four cells), any blastomere can give rise to a complete embryo, including the parts that normally derive from the other blastomeres.

The opposite results obtained by the manipulation of frog and sea urchin embryos seemed to support the contrasting interpretations of the developmental processes defended by Roux and Driesch, respectively. Using the language of philosophy, Roux's vision was based on mechanism, Driesch's instead on a form of vitalism. Roux devoted almost all his life to a research programme called developmental mechanics (*Entwickelungsmechanik*). He was convinced that a sufficiently detailed knowledge of the starting conditions in the egg and of the principles according to which the interactions between the parts (molecules, cells, tissues and organs) take place would have allowed a description of development in agreement with the principles of the physical sciences. According to Driesch, instead, the results obtained from his experiments on the sea urchin demonstrated that in development there is an internal purpose, for which he used the term entelechy, borrowing a word used 2000 years earlier by Aristotle. Driesch also interpreted in terms of entelechy the reconstruction of an animal's missing parts by regeneration.

More than 100 years later, what remains of the contrasting interpretations of development defended by Roux and Driesch, and the conflicting observations on which these were based?

For a long time, comparative embryology has recognized mosaic development and regulatory development as alternative modalities proper to different zoological groups. Today we know, however, that this contrast is not so clear-cut: in the development of most animals there are both regulatory and non-regulatory aspects, although in very different combinations.

The hope that we might one day be able to carry out Roux's project of developmental mechanics has been resurrected in recent decades, although translated into a new language suggested by advances in technology. Today's machine is the computer. The goal of developmental biology, therefore, would be to 'compute' the embryo.

I will discuss in Chapter 6 (p. 97) whether this project makes sense and is achievable. Here, before leaving the line of studies that originated with Roux and Driesch, let's see how far the genealogical history of an embryonic cell within the cell lineage initiated by the first division of the egg can determine the fate of that cell's progeny.

As we can expect from the different behaviour of frog and sea urchin embryos, the situation is very different in the different zoological groups.

It might be a good choice to start with sponges, for two reasons. First, sponges are among the simplest of animals, without real organs and systems: they are not, however, simple clusters of identical cells, but have different cell types with different functions (coating, support, circulation of water and so on). Second, a sponge can be quite easily dissociated into individual cells. The first experiments of this type were performed in the first decade of the twentieth century by the American researcher H. V. Wilson. He managed to separate the cells of a sponge mechanically. Some of the cells thus obtained were destroyed during the coarse treatment, but the others survived and, thanks to their motility and their ability to adhere to each other and to the substratum, showed a notable capacity for reorganization. The resulting cluster of cells, initially loose and irregular, took on an increasingly compact spherical shape. Inside, cells of one type tended to progressively separate from those of other types; this eventually led to a redistribution of cells in a pattern like in a normal sponge. Later experiments confirmed Wilson's results, although some species of sponges showed much greater capacity for reorganization than others.

These experiments demonstrate that, in some organisms at least, differentiation can be reset. The next story will provide an example of the limited reliability of a cell's history along the cell lineage as a predictor of its fate.

In Chapter 1 (p. 16) I mentioned that *Caenorhabditis elegans* disappointed the researchers who expected to find in this worm's embryo a clean example of mosaic development. In *C. elegans*, the position of a cell in the lineage only partly allows prediction of which organs and tissues we will find in its progeny at the end of development. Before morphogenesis begins, many cells move across the embryo, leading to the formation of the groups of contiguous cells that contribute together to the formation of individual organs, but among them there are cells with different degrees of kinship. In other words, more closely

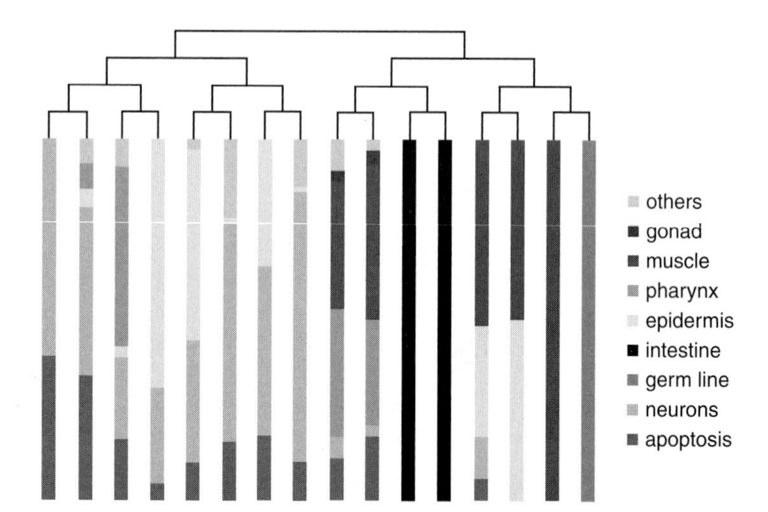

Figure 4.2 Cell lineage and cell fate in *Caenorhabditis elegans*. Starting from the zygote, four runs of cell division produce 16 blastomeres, the progeny of which are differently involved in the production of the different body parts, and a number of cells are fated to programmed cell death. Only four out of the 16 blastomeres include progeny with uniform fate: intestine (two blastomeres), muscle and germ line (one blastomere each); the progeny of the remaining 12 blastomeres has mixed fate. Reciprocally, most body parts include cells issued from a number of distantly related blastomeres.

related cells can be recruited to form different organs, and more distantly related cells can instead contribute with their offspring to the same organ (Figure 4.2).

The Arrow of Individual Life

Development Is Not Always Irreversible

In a short article published in 1909, the zoologist Jovan Hadži (a Slovenian of Serbian origin) reported a very unusual transformation observed in a kind of medusa. To help in putting this observation in its proper context, let us mention that medusae are the free-living, pelagic form of three of the four classes of cnidarians. In the fourth class, the anthozoans, only a sedentary form is present, the polyp, which can be solitary, as in sea anemones, or

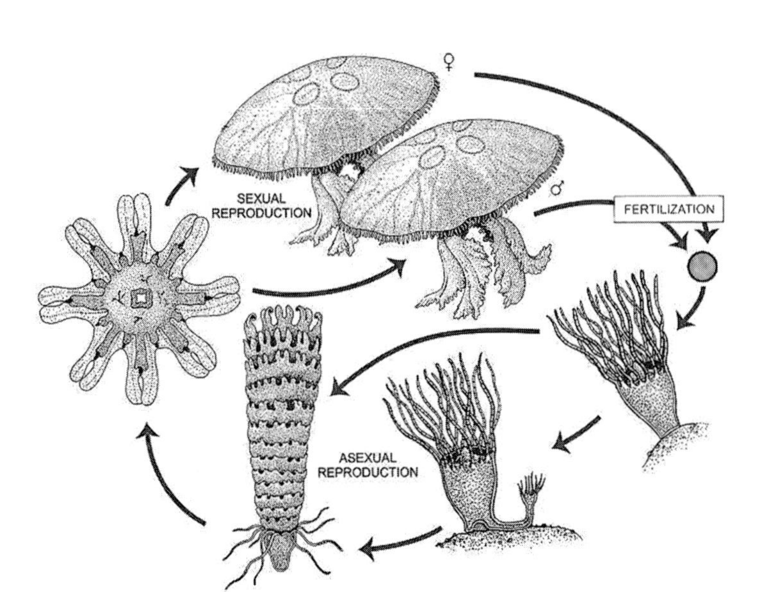

Figure 4.3 In the life cycle of many cnidarians, as in this scyphozoan (*Aurelia aurita*), an asexually reproducing polyp alternates with a sexually reproducing medusa. Reprinted with permission from Fusco, G., and Minelli, A. (2019) *The Biology of Reproduction*. Cambridge: Cambridge University Press.

colonial, as in corals. In the other three classes there are both polyps and medusae, but the relationship between these two forms is not always the same. In cubozoans, a polyp is a larval phase that will metamorphose into a medusa. Of the hydrozoans, some are only medusae, others (like hydra; p. 40) only polyps, but there are many species in which polyp and medusa alternate in the species' biological cycle: the polyps reproduce asexually through buds that develop into medusae that detach from the parent polyp; medusae reproduce sexually, producing a new generation of polyps. In most of the species belonging to the remaining group of cnidarians, the scyphozoans, there is also alternation of generations: from the eggs produced by a medusa, a small polyp is produced, known as the scyphistome, from the upper end of which tiny flat medusae (ephyrae) detach, destined to grow to sometimes very large size (Figure 4.3). *Chrysaora isoscella*, the species on which Hadži made his observations, belongs to the scyphozoans. When keeping ephyrae of this species in an aquarium without supplying them with food, he observed that the small medusae took on a shape similar to the polyp from which they had detached. In other words, Hadži observed a reversal of the normal arrow of the life cycle.

At first sight, this observation called into question a principle that we still tend to regard as indisputable – the irreversible character of development. The possibility of rewinding the tape of the individual's story would open scenarios that seem to belong to the realm of fairy tales. Hypothesizing a return to a phase of individual life that shakes off the ailments of ageing might fuel the illusion of immortality. Indeed, the immortal medusa has been mentioned often in the media, although not at the time of Hadži's observations (practically forgotten since the time they were published), but when a transition from medusa to polyp was observed again in 1988 by Christian Sommer, a young German student. This discovery was published only in 1992 in the first of a series of papers, mainly co-authored by researchers from the Universities of Genoa and Salento (Giorgio Bavestrello, Stefano Piraino and Ferdinando Boero), who in the years since then have clarified many aspects of the complex developmental biology of this animal. This second cnidarian capable of medusa-to-polyp transition was a hydrozoan called *Turritopsis dohrnii* (in the initial publications, a different name, *Turritopsis nutricula*, was used). In *Turritopsis*, a medusa can reverse the arrow of development at any time of its existence, under adverse conditions such as lack of nourishment or sudden changes in temperature or salinity.

It can be disputed how far the behaviour observed in this species (or in Hadži's *Chrysoaora*) actually deserves to be called reverse development. It does not, if examined in the context of the animal's life cycle. In fact, if the medusa-to-polyp transition observed in *Chrysoaora* or *Turritopsis* is a developmental event, the reciprocal polyp-to-medusa transition simply does not exist as a developmental event, because polyps (except for those of cubozoans, a group in which no case of 'reversal' is known) do not change into medusa, but give rise to medusa by asexual reproduction. Therefore, the peculiar events observed in *Chrysaora* and *Turritopsis* are an instance of reversed development only in terms of genetic generations (same genet), but not of demographic generations (distinct ramets); the relevant concepts are discussed on p. 33 (genet versus ramet) and p. 85: genetic versus demographic generation).

However, true instances of reversed development are found as soon as we explore the entire group of cnidarians more carefully: in some anthozoans, young polyps can return to a stage of planktonic larva.

In a sense, a transitory phase of reverse development also occurs: what I suggested in 2003 should be called 'Lazarus' developmental features. Why Lazarus? Because, like the friend of Jesus who returned to life, these traits, manifested at an early stage of development, seem later to disappear but end up reappearing further on. Similarly, palaeontologists have been talking for some time of Lazarus taxa, to indicate those forms that are known from fossils of two different ages, sometimes widely spaced, but not in rocks of intermediate age: the temporal continuity of the lineage is not documented by the fossil record, it can only be inferred.

Mites offer an example of a Lazarus developmental feature: in the embryonic stage, many of them have primordia of four pairs of legs, but the larva has only three pairs: the fourth will show up again in the nymph and will still be present in the adult. Appendages that disappear and then reappear in the course of post-embryonic development are also observed in some species of shrimp.

More intriguing is the case of the genital appendages (gonopods) of the males of some millipede species. Here, an individual that has reached sexual maturity can still undergo one further moult. However, the stage following this moult is not always a reproductive stage like the previous one. In some species, between two successive mature stages is an intercalary stage in which the genital appendages have regressed; these will form again only with the next moult, after which the condition of sexual maturity is restored. The transition from a mature to an intercalary stage bears some resemblance to the reverse development of the 'immortal' medusa: this transformation too is induced by adverse environmental conditions and ends up prolonging the life of the animal.

Organogenesis?

Individual Organs Are Not the Products of a Distinct Developmental Process

As we have seen several times in the previous pages, the questions we ask about development depend very much on a basic choice: are we interested in the history of the individual (for example, the development of a human or a fruit fly from zygote to adult), or in developmental processes characterized by specific mechanisms? When exploring embryonic development starting from the zygote, we encounter a phase, called organogenesis in embryology textbooks, about which we need to be critical precisely in light of this double meaning of the word development.

According to traditional descriptive embryology, three main phases follow the fertilization of the egg – we are focusing here on animals, of course. The subdivision of embryonic development into cleavage, gastrulation and organogenesis, which is satisfactorily applicable to most zoological groups, although with exceptions, remained in use even when the strictly descriptive approach to embryology was followed by an experimental approach through mechanical manipulation or chemical treatment of embryos, and eventually by the molecular genetic approach that characterizes research today.

However, it is necessary to ask how and to what extent the phases (cleavage, gastrulation and organogenesis) we recognize at morphological level correspond to distinct processes and mechanisms. In other terms, granted that organogenesis is a stage of development, can we say that it is also a developmental process?

It is easy to expect a negative answer to this question, if only for the growing diversity and complexity of the structures that emerge during organogenesis. Contrary to widely shared opinion, I think that only in very generic (and therefore uninteresting) terms can we equate building a heart to building a brain. So let's immediately go down to a more detailed scale and ask if there are specific mechanisms or processes that characterize the construction, for example, of a heart.

The issue can be addressed at different levels, for example genes or cellular mechanisms. We will delve into the first aspect in Chapter 5 (p. 100), where I will discuss the controversial notion of master control genes. This name has been given to a small number of 'powerful' genes, each of which is believed to be primarily responsible for controlling the construction of a body part: the heart, for example, would have its own master control gene. Let us examine here briefly the possible existence of organ-specific processes at the cellular level, responsible, for example, for the construction of a heart or a brain. Once more, my answer is negative.

Even structures that seem particularly homogeneous are often built by assembling parts together, the construction of which started independently and by different mechanisms. In animals whose body is divided into segments (more on this in Chapter 7, p. 116), different parts of the body are often made of segments of different origin. In some marine annelids, a small number of segments are formed very early and are already recognizable in the larva; the remaining segments are added after metamorphosis. Many scholars also acknowledge in arthropods a distinct origin for the anterior segments (those of the anterior half of the head) with respect to the following ones.

In insects, the respiratory system is made up of tubes (tracheae) that open on the surface of the body through spiracles and send branches to the different tissues and organs: in these animals, the distribution of gases is not entrusted to the circulatory system, but is accomplished directly through the tracheal system. The thinnest branches of the latter are made up of very thin intracellular tubes, called tracheoles. The origin of tracheoles is different and separate from the tracheae, which are invaginations of the epidermis (and, like the latter, are covered with cuticle). Therefore, an insect's respiratory system is formed by assembling two kinds of elements (tracheae and tracheoles) that are first formed independently.

In the vertebrate nervous system, the cells of the spinal cord have a separate origin from those that make up the brain. Even more diverse is the origin of the cells that form the heart of a vertebrate. Two main contributions are provided by local mesodermal cells, which differentiate to form the endocardium (the tissue layer that lines the heart chambers) and the heart muscle, and by other progenitors that supply the cells of the epicardium (the layer of connective tissue immediately outside the heart muscle) and coronary vessels; to these are added cells immigrating from afar, which form the lymphatic system of the heart. But there is more to say. If the cells that form the heart have different origins, it is also true that the intercellular signals involved in the development of the heart are not specific to cardiac morphogenesis but are widely used in different embryonic processes.

The assembly of initially separate parts may seem more suited to an industrial plant than to a developing organism, but there are even more sensational examples than those mentioned thus far. The most remarkable case is provided by nemertines. These worm-like animals (their habitus justifies their common name of ribbon worms) are almost all marine. We have already met them in Chapter 1 (p. 8), where I mentioned the very strong negative growth they are capable of. Most of these worms have indirect development, and their larvae are varied, but often very different from adults. The metamorphoses they undergo are among the most dramatic in the whole animal kingdom. Most remarkable are those of the nemertine larva known as a pilidium. Eight separate cell groups differentiate in these larvae, from which the future adult will take shape. A fundamental step in this metamorphosis is the gathering of these groups of cells, called rudiments, to form a single body. A pilidium is truly a mounting kit, but the instructions to be followed to make an adult ribbon worm are still almost unknown.

In summary, organogenesis is not a developmental process, but only a name for the developmental phase in which the main parts of the body take shape, through an interweaving of different processes.

Critical Stages

Differences Between Species Do Not Increase Progressively from the Egg Onwards

In the first half of the nineteenth century, embryology was a merely descriptive science. In 1828, the year in which Karl Ernst von Baer published the first volume of his treatise *Entwickelungsgeschichte der Thiere – Beobachtung und Reflexion* (Animal Embryology – Observations and Discussion), the species in which embryonic development had been studied were still very few, and largely restricted to vertebrates. Nevertheless, von Baer ventured into comparisons from which he derived a 'law', divided into four points, as follows:

1. What is common to a larger group of animals forms earlier in the embryo than what is special to a smaller group or a species.
2. Less-general shape relationships progressively emerge from more general ones, until finally the most special occurs.
3. Each embryo of a certain animal, rather than passing through the forms specific to other animals, becomes increasingly different from them.
4. Basically, therefore, the embryo of a higher animal is never similar to the adult of another animal, but only to the latter's embryo.

Among the authors who ventured into this type of comparison in the following decades, the best known is the German zoologist Ernst Haeckel. The first work in which he wrote about this topic was published in 1866 under the impressive title *Generelle Morphologie der Organismen* (General Morphology of Organisms). Seven years before, Darwin had published his momentous book *On the Origin of Species*, so it is no wonder that Haeckel expressed his principle in evolutionary terms: in his words, ontogeny (the development of the individual) is a short recapitulation of phylogeny (the evolutionary history of the species).

In the end, that the differences between two animals become progressively more evident as their development proceeds (Figure 4.4) is just what we should expect. In the earliest stages, embryos are made of just a few cells, which can only be arranged in a limited number of different reciprocal positions; and, above all, the number of these cells is too small to make complex and varied structures such as the different parts of the body of an adult animal.

If we also take into account that all animal species derive from a distant unicellular ancestor, it is easy to understand the success of a model that represents the progressive divergence of the developmental trajectories of different animals in the form of a funnel, starting from a common origin.

However, subsequent advances in descriptive embryology have revealed that the funnel model does not correctly represent the earliest stages of development. These, in fact, can be very different even among closely related species. A convincing example is given by two species of sea urchins, on which a considerable research effort has been focused by the American (Canadian-born) biologist Rudy Raff and his collaborators, in particularly the American biologist Greg Wray in the last years of the twentieth century. These two species of sea urchin are so similar to each other that they are classified in the same genus. Their similarity, however, is manifested only in the full-grown animals, while their earliest developmental stages could not be more different. One of the two species, *Heliocidaris tuberculata*, produces small eggs (95 micrometres in diameter), with a very modest provision of yolk. The embryonic development through the usual steps (cleavage, gastrulation and so forth) leads to the formation of a planktonic pluteus larva, which needs to feed before undertaking metamorphosis. The other species, *Heliocidaris*

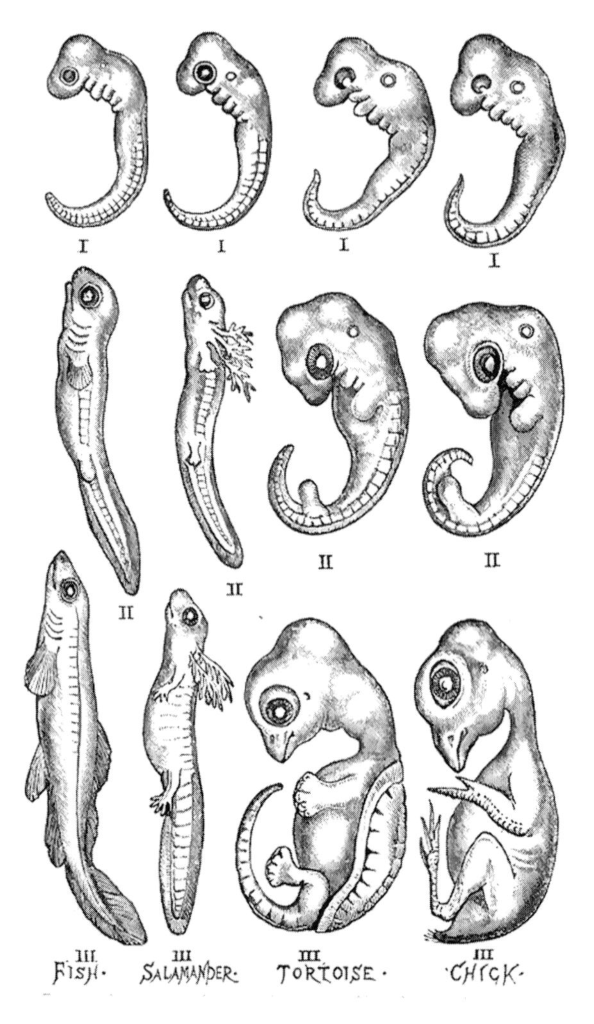

Figure 4.4 The plates in these two pages are the most popular version of the comparison, first proposed by Ernst Haeckel in 1874, between equivalent developmental stages in different vertebrates.

This version is reproduced from George John Romanes' (1892) popular book on Darwinism.

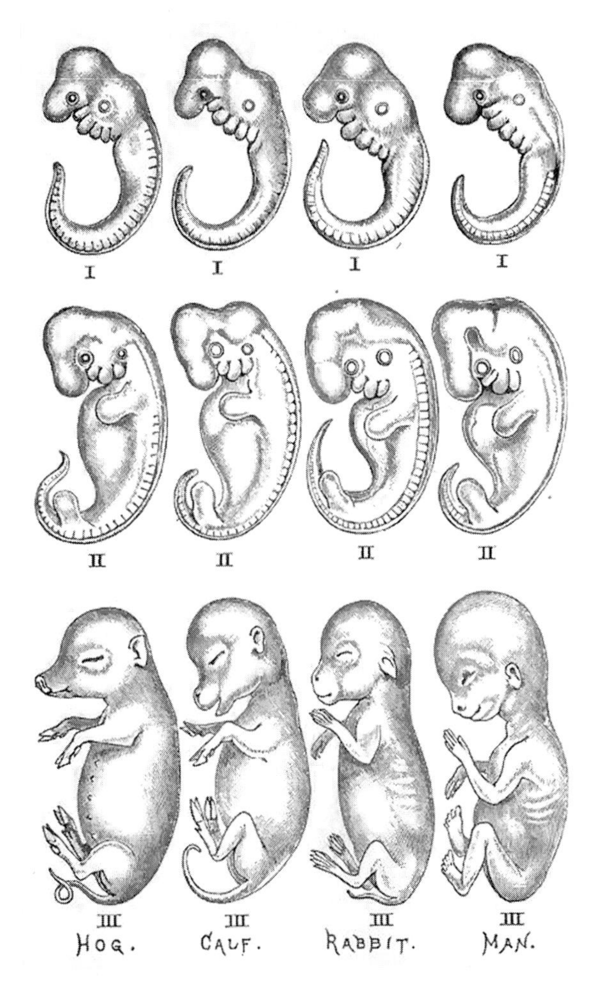

Figure 4.4 *(cont.)*

erythrogramma instead produces large eggs (400 micrometres in diameter), with abundant yolk. Because of this, the usual embryonic stages are skipped, and the still yolk-filled embryo develops directly (that is, without larval stage and metamorphosis) into a small sea urchin. Despite these dramatic differences in early development, the two species are indeed very closely related, as confirmed by the hybrids between *H. tuberculata* and *H. erythrogramma* obtained in the laboratory.

The stage at which the two species are most similar, therefore, is not at the beginning of development (in the egg), but later. The developmental trajectories of the two species converge initially towards this stage, but diverge again from that point on. A similar pattern is observed in other zoological groups, for example in insects and vertebrates. In 1983, the German embryologist Klaus Sander introduced the term *phylotypic stage*, to indicate this point at which the species of a given zoological group are most alike. In 1994, the Swiss-French developmental geneticist Denis Duboule introduced the hourglass metaphor to describe the trend of differences between species of the same group during embryonic development: the initial differences gradually decrease up to the phylotypic stage (the narrow neck of the hourglass), then begin to diverge.

According to some researchers, such as the British embryologist Michael Richardson, it would be more appropriate to speak of a phylotypic period, rather than a phylotypic stage. In any case, the nineteenth-century model of the funnel is definitely inadequate.

In the embryonic development of large zoological groups such as vertebrates and arthropods, it is thus clear that the progressive manifestation of the peculiarities of each species does not begin with the division of the zygote into two blastomeres or, in any case, at a very early stage of development, but only later. This is in agreement with two properties of eggs. First, the egg is far from being the least specialized of cells, starting from which a history of progressive specialization of daughter cells begins, causing an increase in the complexity of the embryo (p. 91). Second, mRNA and proteins that, although reflecting the maternal rather than the zygotic genome, regulate the whole morphogenesis of the embryo in the initial stages are dismantled before reaching the phylotypic stage (p. 33).

However, generalizations must be avoided. In the history of an individual's development, it is hard to identify a phylotypic stage if the animal does not derive from an egg, but from a bud or a fragment of the parent's body. Furthermore, a phylotypic stage seems to be missing in the sponges: this should not be a surprise, in animals capable of reorganizing themselves from a cloud of dissociated cells (p. 63).

But there is more. A recent study on two mollusc species (oyster and abalone) and a marine annelid indicates that in these zoological groups the embryonic development follows neither the funnel nor the hourglass model. By comparing the genes transcribed by these animals at different times through embryonic development, it was found that differences between species are highest at an intermediate stage of embryonic development, precisely when we would expect the greatest similarity characteristic of the phylotypic stage. Before and after this phase of divergence, the overall differences in gene expression between species are distinctly lower. To sum up, two centuries since the formulation of von Baer's law, we can still expect interesting discoveries from future research effort.

5 Developmental Sequences: Sustainability versus Adaptation

The Hydra and the Colonial Sea Squirt

Development Does Not Always Begin with the Egg

The first to describe the freshwater polyp we know as the hydra was the Dutch businessman and self-taught scientist Antoni van Leeuwenhoek, in a letter written on Christmas Day of 1702 and published on 1 January of the following year in the *Philosophical Transactions of the Royal Society* of London. In it, he described the polyp's ability to reproduce by budding.

As mentioned in the first chapter, reproduction and development are two inextricably intertwined aspects of biology. Thus, with the few paragraphs dedicated to the tiny freshwater polyp, Leeuwenhoeck offered the first description of asexual reproduction in this animal as well as an indication that, in order to build an individual, it is not necessary to start from an egg.

Asexual reproduction and development of an individual from somatic cells are now known for many animals, in which these processes are often an alternative to sexual reproduction and the embryonic development that follows it. Some populations (and some laboratory strains, for example of planarians) have only asexual reproduction, but the new individual is not always produced by means of buds that detach from the body of the parent individual, leaving the latter practically identical to what it was before producing the bud. In a number of annelids and flatworms, asexual reproduction occurs through the fragmentation of the parent organism into several parts, each of which gives rise to a new complete individual. In this case, the process is not very different from regeneration, apart from the fact that no trauma is involved. I will deal with regeneration in another section of this

Chapter. Here, I focus on the differences between development from an egg through embryonic stages and development from a bud, in animals in which both modalities exist.

We could start from hydras, which reproduce almost always asexually, by means of buds, but are also capable of sexual reproduction, with the production of gametes and the formation of zygotes from which embryonic development starts, leading to the production of new individuals. Between a hydra produced by budding and a hydra generated by sexual means there do not seem to be any anatomical differences, but this is not surprising given the very simple structure of these polyps.

However, we could start, instead, from a different and more interesting point, namely a comparison between sexually and asexually generated individuals in a group of morphologically more complex animals in which both reproductive modalities occur in the same species (indeed, these alternate regularly in the biological cycle). Some of these animals have been the subject of detailed research. We therefore move on to the ascidians, in particular to the colonial ascidians of the genus *Botryllus*. Ascidians or sea squirts belong to the tunicates, a group of marine animals that get this name from the often conspicuously coloured tunic that covers their body, which has two large openings, the atrial syphon from which water enters and the cloacal syphon from which water exits, leaving on the large gills the suspended particles on which the ascidian feeds. Despite their very different adult organization, ascidians and the other tunicates are ascribed to the same phylum as vertebrates (Chordata), because they share the possession, at least in embryonic or larval condition, of a dorsal rodlike structure called the notochord.

A tiny larva vaguely similar to a tadpole originates from the fertilized egg of the ascidian. The larva moves in the water for some time, then settles on a substratum, undergoes metamorphosis and eventually grows to adult. This short description applies to all ascidians. However, several species can also reproduce asexually by different modes of budding. In *Botryllus schlosseri* (and others) this ability manifests itself very soon. A bud or two are already visible in the tiny larva when it leaves the parent's body; following its attachment to the substratum, it will quickly originate an entire colony. The units that make it up (individuals permanently joined together through the blood vessels, whose system is shared by the whole colony) are referred to as zooids.

Even before completing their development, the zooids generated by the buds produced by the larva give off a second generation of buds, and the zooids into which these develop produce, in their turn, third-generation buds. Three successive generations of zooids are usually present in a colony of *Botryllus*, but its membership – up to a few thousand zooids – is constantly renewed. In the course of a week, the zooids of the older generation degenerate, replaced by those of the next generation, and so on for the younger zooids. Only after a few cycles of asexual reproduction do zooids with mature gonads show up. In some species of colonial ascidians, asexual reproduction is even more impressive: a new zooid can originate even from a single blood cell. In *Botryllus* this extreme form of asexual reproduction does not occur in nature, but it can be induced by removing all the zooids from a colony, which is thus reduced to the network of blood vessels and their content.

In *Botryllus* we thus have the opportunity to compare the two different developmental processes by which a zooid can be formed, starting from a fertilized egg or from a bud, respectively. In the first case, the animal develops through the usual embryonic stages, followed by larval stage and metamorphosis; in the second case, there is no embryonic development (therefore, no blastula, gastrula and so on) and no larval stage. Differences become even more striking when we realize that in the formation of a zooid from a bud there is no differentiation into the usual germ layers: ectoderm, mesoderm and endoderm. In the zooid that originates from the egg, on the other hand, germ layers are differentiated, and the cells derived from each of them will have the expected fate: epidermis and nervous system, for example, will derive from the ectoderm; musculature, circulatory apparatus and blood cells from the mesoderm; intestine from the endoderm. Morphogenesis is very different when a zooid takes form from a bud, despite the similarity of processes at the cellular and molecular level. The latter can be described as a double vesicle: the outer layer, formed by the epidermis of the parent zooid, encloses the internal vesicle, which in *Botryllus* is of ectodermal origin, while in the zooids formed from blood cells (in other ascidians) it is of mesodermal origin. Whatever its origin, the internal vesicle produces most of the organs, including intestine and nervous system.

These dramatic differences between the two modes of development, however, tend to disappear as organ differentiation progresses. In the end, between a zooid that derives from the egg and is formed through embryonic and larval development and a zooid deriving from a bud, there is practically no difference, apart from the fact that the former does not reproduce sexually. An ascidian specialist might remark that the meshes of the animal's large gill are somewhat different in the two types of zooids. Other than that, they are practically identical, even at the level of the cell types that make them up.

Mature versus Adult

Animals and Plants Do Not Necessarily Reproduce as Adults

Ctenophores or comb jellies are marine animals with a diaphanous body. Most of the known species lead an existence similar to that of a jellyfish. Like those, they are predators, catching prey thanks to the sticky secretion of peculiar cells, the colloblasts. Very diverse in shape and size, from a few millimetres to over a metre, all ctenophores have eight rows of 'combs' formed by hundreds of agglutinated cilia, which allow some manoeuvring in the water. We are interested here in these animals because the reproductive behaviour of some species of ctenophores challenges what we usually mean as 'adult'.

Like other terms in biology, 'adult' is generally understood in the sense in which the term applies to the human species and, more generally, to the animals with which we are most familiar, such as mammals or birds. Adult is therefore the individual at the end of development, when it has reached sexual maturity and – as we often say – the traits proper to its species. I will not insist here on criticizing the overt adultocentrism of the last sentence: I have already discussed the topic in Chapter 1 (p. 4). I will just mention that in some zoological groups it is easier to identify the species to which a specimen belongs from a larva rather than an adult. But let us see to what extent the achievement of the adult condition actually aligns with the beginning of reproductive activity. It is in this respect that ctenophores have something interesting to show.

In some ctenophore species, young individuals, far below the size they will eventually reach later, begin to produce gametes. As a rule, this first breeding season soon ends, and the animal starts growing again, eventually reaching maximum size. Now, in a condition we can at last call adult without hesitation, it starts producing gametes again, only stopping if it is running out of food. If this happens, the animal ceases to invest in reproduction and begins to shrink – this is an example of negative growth, the process briefly discussed in Chapter 1 (p. 8). Regarding growth, ctenophores are somewhat plastic: if food is available again, they resume growing and eventually producing eggs and sperm.

In a sense, ctenophores show a sort of prolonged maturity, distributed over a segment of life and a range of sizes of considerable extension and not precisely fixed. Some species of ctenophores, however, go far beyond that. They reproduce in two moments of individual life that differ not only in age (and size), but also in body structure. In this phenomenon, known as dissogony, a first reproductive phase occurs when the animal is still a larva. This does not last long. Reproductive activity will be resumed much later, after metamorphosis and a subsequent period of growth. Ctenophores of the genus *Mertensia*, for example, have a first reproductive phase as sexually mature larvae just 1.5 millimetres long, whereas the conventional adult (final) stage, in which the reproductive capacity is reactivated, measures up to 10 centimetres. But the story does not end there. In the central basin of the Baltic Sea lives a population of *Mertensia ovum* that includes only larvae that do not undergo metamorphosis any longer. Were the behaviour of the other ctenophores not known, including dissogony in other populations of *M. ovum*, we would not hesitate to qualify the tiny Baltic *Mertensia* as adults. Instead, even if they do not undergo metamorphosis, we continue to call these reproducing individuals 'larvae' – a decision that we should perhaps revise.

A double breeding season comparable to the dissogony of ctenophores is almost unknown in other animals, although in those that have a life long enough not to exhaust their reproductive capacities in a single season, the moments in which the individual is sexually mature may be separated by long intervals. Let us think, for example, of the many terrestrial vertebrate species of the temperate regions that have a reproductive season every year of their life. In many arthropods, the individual that has reached sexual maturity may still undergo one or more moults: in this case, the passage from one instar to the next is accompanied by a fairly long break in the reproductive season. In addition to the millipedes mentioned in Chapter 4 (p. 67), multiple sexually mature stages separated by moults are common in crabs and in other groups of arthropods, winged insects excluded. Among the latter, the only group that approaches this condition is the mayflies, if only in a virtual way. Most mayfly species have two winged stages separated by a moult, but the first of these stages (called the subimago) lasts only a few hours and is not involved in reproduction: this is reserved for the final stage (the imago), which will survive for just as short a time.

There are other interesting stories about insects. In some species, the adult is able to reproduce immediately after the moult, in others instead it can bring the gametes to maturity only after properly feeding. The latter is the condition of female mosquitoes, whose sexual maturation depends on a good blood meal. However, there are also those that play in advance. In the *Paratanytarsus grimmi* gnat, for example, the adult is already sexually mature when it is still wrapped in the cuticle of the pupa, the resting stage between larva and adult. And there are species that anticipate sexual maturity much more, to the point of approaching what happens in the Baltic ctenophores. In several species of gall midges (tiny dipterans of the cecidomyiid family), maturity is reached at the pupal or even the larval stage.

Generational Problems

Development Does Not Always Start from Scratch with the New Generation

In 1997, the Portuguese developmental biologist Clara Pinto Correia published a book entitled *The Ovary of Eve: Egg and Sperm and Preformation*, a very well documented and pleasant work to read. Why the ovary of Eve? Let us first introduce a theory quite popular among scientists and philosophers through the second half of the seventeenth century and for most of the following century. This theory is commonly known as preformation, and Pinto Correia follows this usage. However, the British philosopher Andrew Pyle has compellingly argued that preformation means that the new individual is somehow elaborated in the ovary of the mother or in the testes of the father; thus the term should really be reserved for a theory defended, for example, by Italian physician and philosopher Fortunio Liceti in *De spontaneo viventium ortu* (On the Spontaneous Origin of Living Beings; 1618). A different term, pre-existence, should be used instead for the theory of development we are about to outline here.

In the seventeenth century, an increasing body of observations through the lenses of the first compound microscopes was revealing the existence of a world of living beings invisible to the naked eye. That those tiny bodies were living things was not in doubt, at least in cases where the microscope showed their ability to move, sometimes so quickly that they escaped from the small visual field allowed by the lenses before the observer could determine their shape accurately. Some microscopists observed what would later be called protozoans, others – more intriguingly – the *animalcula* (little animals) present in semen. Sperm cells were believed by some early microscopists to be parasites, whereas others interpreted them as the active vital element contributed by the male parent. At the time, it was not accepted by all that father and mother contributed in equivalent degree to the formation of the embryo. In the case of mammals, the egg of which was still unknown, the discovery of the 'little animals' in the semen strengthened the idea of the pre-eminent role of the father, to the point that the maternal contribution was believed to be limited to the nutrition and protection of the embryo. Some microscopists even reported having seen a miniature man (*homunculus*), in what we now call the head of a human sperm.

This is the animalculist version of the pre-existence theory. Others, however, preferred to believe that the future individual was already present in the egg, and are therefore known as the ovists. This explains the title of Pinto Correia's book. The title must be taken literally. If a miniature copy of the future offspring is present in the mother's body, why rule out that this tiny creature might already bear in herself an even smaller copy of what will be her descendants? Extrapolating this nesting of successive generations further back in time, we end up imagining that all future humanity was already outlined in the ovary of Eve, in the manner of the Russian dolls.

This idea, first suggested in 1674–5 by the French philosopher Nicolas Malebranche, was supported by other authors for over a century. Eventually, in his *Considérations sur les corps organisés* (Considerations about Organized Bodies; 1762), the Swiss naturalist and philosopher Charles Bonnet discussed the disquieting implications of this theory. According to Bonnet, the number of future generations cannot be infinite, because this would require an indefinite divisibility of matter to accommodate all generations, one inside the other, in the ovary of Eve. However, by downsizing the idea of nesting to a large but finite rather than infinite number of generations, it was possible to find some situations in nature that seemed to agree with this model. For example, as noted by Bonnet, up to four successive generations of the plant can be observed in a hyacinth bulb. Of course, these boxed plant generations are produced asexually, and thus no female reproductive organs are involved – that is, no ovary.

It might have been more difficult to find examples of boxed generations in the animal world. However, at least one was available from the studies of Bonnet himself, who had observed unmated females of aphids (plant lice) giving birth to individuals similar to them except for their smaller size. It was later discovered that this offspring is produced by parthenogenesis (reproduction through unfertilized eggs) associated with viviparity. With aphids, it was not difficult to imagine that the offspring was already outlined in the mother's body long before birth. In fact, in the ovaries of parthenogenetic female aphids there are the embryos of their daughters, and within the ovaries of the latter, the embryos of the daughters of their daughters.

A similar arrangement occurs in a genus of tiny flatworms (*Gyrodactylus*), external parasites that live on the fins or skin of freshwater fishes. In these worms, which are hermaphrodite, the egg is generally fertilized by a spermatozoon produced by another individual, but self-fertilization can also occur. In any case, the fertilized egg (only one at a time) starts developing inside the parent's uterus, but before the embryo is released, a second embryo is already produced inside it; within this second embryo develops a third embryo, and finally a fourth within the third. When the first embryo is released by the parent, the previously arrested development of the second embryo is reactivated, and so on.

In this description of nested embryos, we have been speaking of successive generations. Even before the new individual begins its autonomous life, it does not seem difficult to establish a dividing line between the generation to which it belongs and the generation of its parent. It might therefore seem that establishing the boundaries between one generation and the next is always possible and not arbitrary. But this would be a hasty conclusion.

Let's leave aside the small parasitic worms with nested generations, to look closely instead at the development of better-known animals, for example the fruit fly.

From a certain point of view, a new generation of *Drosophila* begins when an egg is fertilized and a zygote with a new genome (the zygotic genome) is obtained. The latter is the sum of maternal and paternal contributions.

However, for some time the zygotic genome is not expressed. This is particularly evident in animals such as *Drosophila*, in which the embryonic cell divisions proceed at so fast a rate that there is no time for the transcription of the zygotic genome. In the meantime, only copies of mRNA and proteins matching the maternal genome are available in the embryo's cells. Finally, when the rapid initial sequence of mitotic divisions slows down or is temporarily stopped, the zygotic genome begins to be expressed.

As discussed before (p. 40), one of the criteria for recognizing individuals is the possession of an exclusive genome, but it seems difficult to apply it in a situation where cells with a given genome (in the nucleus) behave under the control of a different (maternal) genome, important traces of the expression of which are present in the cytoplasm. So long as the embryo has not undergone the so-called maternal–zygotic transition, it seems appropriate to recognize an overlap between generations. The new generation will eventually emerge from this ambiguous situation only when the zygotic genome finally starts to be expressed, while the remaining mRNAs of maternal origin are actively destroyed.

To better understand the meaning we can give to the word generation, let us take a look at unicellular organisms and start with purely demographic considerations – problems of counting individuals.

Consider a unicellular organism, for example a bacterium or a ciliate that multiplies by symmetric division. We may say that the initial F_0 cell is totally replaced by two F_1 daughter cells. But this means that no F_0 individual survives with its offspring, a condition very different from the generational turnover in animals or plants. In what sense, then, do two bacteria or protozoa issued from a 'founding' cell represent a new generation? One might be tempted to equate the clone of unicellular organisms whose membership grows with each cell division to the clone of blastomeres issued by cell divisions in the embryo. But in a clone of bacteria or protozoa, cells, although genetically identical as are blastomeres in the embryo, are also physically distinct: it is possible to count them and therefore to treat them as physically distinct individuals as we do with multicellulars belonging to the parental versus filial generations. Moreover, in those weird unicellular organisms the ciliates, whose unique biology again deserves attention in this book, there are additional complications.

We know from Chapter 2 (p. 31) that in ciliates there are phenomena of sexuality, but these have nothing to do with reproduction. Through a process called conjugation, two ciliates reciprocally exchange a haploid copy of their genome but do not merge to form a zygote. Once this exchange is done, the two ex-conjugants separate. Let us make some simple calculations. Before conjugation there were two individuals, and after conjugation there are also two. So there has been no reproduction. Is it therefore appropriate to attribute the ex-conjugants to the same generation as the two individuals who took part in the conjugation? In a demographic sense, the ex-conjugants do not represent a new generation. However, from a genetic point of view they differ from the two conjugants in the same way as the progeny of a sexually reproducing animal differ from their parents. It may therefore be appropriate to break down the traditional concept of generation into two separate concepts: demographic generation and genetic generation. In a demographic sense, the passage from one generation to the next is marked by reproduction; in a genetic sense, the passage to the next generation is marked by genetic changes, due to sexuality.

In sexual reproduction, a demographic generation is also a genetic generation, and vice versa. In asexual reproduction, however, only demographic generations can be recognized, within the same genetic generation. In the case of ciliate conjugation, which is a purely sexual phenomenon, the ex-conjugants belong to a new genetic generation but to the same demographic generation as the former conjugants.

There is more to this issue. If we accept a definition of generation based only on genetic changes, we should also acknowledge the beginning of a new generation when meiosis occurs, with the transition from diploid to haploid (often accompanied by genetic recombination). The same is true at the time of fertilization, with the reconstitution of the diploid condition and, as a rule, the formation of a zygotic genome that did not exist before. But this means, in animals for example, that gametes are a genetic generation other than that of the individuals that produce them. It is an unconventional idea, but it may be useful to explore its implications.

Starting Again

Why Should Regeneration Be 'One of the Great Mysteries of Biology'?

We have seen in Chapter 4 (p. 60) how the regulatory capacity of sea urchin embryos made Hans Driesch suspect the presence within the organisms of a principle of intrinsic purpose, capable of manifesting itself even in conditions other than normal. One full century later, it is unlikely that biologists will still defend the vitalistic positions of the German scholar, but there is a whole class of developmental phenomena whose interpretation seems to many researchers difficult to frame in the current, more or less reductionist schemes of developmental biology.

Despite remarkable progress in the study of regeneration in the last decades, as recently as 2018 the American biologist Peter W. Reddien, one of the world experts in regeneration, wrote that "the ability to regenerate missing body parts is one of the great mysteries of biology." In terms of the actual processes involved at the cellular and molecular level, this is much less true today than it was in Driesch's time, and it is fair to add that Reddien has contributed to progress with his important studies on planarian regeneration. But, as a matter of principle, I also do not see any reason to characterize regeneration as a mystery, unless we are willing to say the same of development (or life) in general. If we adopt an inertial view of developmental processes, regeneration is 'just' one of the expressions of the developmental capacity of living systems, on a par with embryogenesis or with the development of asexual progeny via buds. This does not mean that these three processes (embryogenesis, regeneration and development from a bud or a fragment) are simple reiterations of an identical process; I would say, instead, that the fact that largely similar individuals are produced (or restored) through this diversity of developmental pathways denies their putative character of exceptionality, or mystery.

Regeneration often involves the production of a well-defined body part, such as the tail or a leg, that grows to replace a lost equivalent following accident or experimental manipulation. Sometimes, more conspicuously, a complete individual is produced starting from a small bit of the original one, a performance of which very few animals are capable. Piecemeal long-term regeneration, however, can go unnoticed and yet involve complete renewal of the body. This happens in some polyps, such as *Hydra vulgaris*, and planarians, such as *Schmidtea mediterranea*. In the former, all cells of the body column are renewed every 3 weeks, and those forming the tentacles survive for just 4 days. The lost cells are replaced through the sustained activity of a large pool of stem cells. The same has been observed in *Schmidtea*.

We might be tempted to generalize, concluding that the lack of comparable abilities of self-renewal in animals such as a fly or a mammal reveals the much simpler structure of hydras and planarians, compared with insects and vertebrates. However, besides the fact that planarians are anatomically much more complex than hydras, and in view of the lack of self-renewal power in animals such as nematodes, which are not more complex than planarians, it is also fair to add that even among closest relatives there can be large differences in seemingly 'generic' properties such as the attitude to self-renewal. An example is provided by hydras. It has been estimated that in suitable, constant environmental conditions, the efficient self-renewal of *Hydra vulgaris* would allow a polyp to live 1400 years at least: this is not immortality, but a generous effort in that direction. This performance, however, is not shared by the polyps of *Hydra oligactis*, in which, despite the activity of stem cells, signs of senescence show up before long.

Two centuries before these important details of cell renewal were discovered, *Hydra* had already found a place among choice models for the study of regeneration. When he found these small polyps in a canal in The Hague, in 1740, the young Swiss naturalist Abraham Trembley did not know of Leewenhoeck's previous description, mentioned in Chapter 5 (p. 76). Some of the hydras collected by Trembley were green, a peculiarity which, added to the ability to produce buds, seemed to demonstrate their belonging to the plant kingdom. (About 1880, two German researchers, Carl Semper and Karl Brandt, would demonstrate that the colour of green hydras is due to the presence of endosymbiotic green algae.) However, their soft body was able to contract if stimulated – an argument in favour of their animality. This ambiguity prompted Trembley to undertake a long series of observations and delicate experiments that, in addition to demonstrating his ingenuity and handling ability, revealed in the small polyp regenerative capacities far beyond anything known until then (and most of what subsequent experiments have been able to reveal). Rather than solving the dilemma about the nature of the hydra, either animal or plant, the results of Trembley's experiments seemed to have an equivalent only in myth, in the swamp monster known as the Hydra of Lerna, in which a severed head (it had nine) would immediately form again.

At the level of the type and behaviour of the cells involved, it is advisable to distinguish two types of regeneration, called epimorphosis and morphallaxis respectively.

Epimorphosis involves cell proliferation, with the formation of new tissues: the first step is the closure of the wound, followed by recruitment of initially unspecialized cells (stem cells) that form the small mass (blastema) destined to grow and differentiate into the replacement part; morphallaxis is the remodelling of the tissues close to the wound. Interestingly, this distinction was first proposed by Thomas Hunt Morgan, who spent several years studying regeneration before becoming one of the main figures in genetics (I mentioned him on p. 15, in respect to the famous Fly Room at Columbia University).

Probably no animal can compete with hydras or planarians in regenerative capacity. Both have a large amount of stem cells from which all cell types required to complete a missing body part can be obtained: it has been estimated that stem cells represent between 5% and 10% of all cells in an adult planarian. These are the only cells capable of dividing: all differentiated structures of the animal are made up of cells that at an early stage of development have lost the ability to divide.

As for hydras, a study on the cellular composition of *Hydra attenuata* published in 1973 gave an estimate of about 60 000 cells per polyp, tentacles excluded. In these animals, a fragment of a few hundred cells is sufficient for the regeneration of a complete polyp.

There is no relationship between an animal's size and complexity on the one hand and its regenerative abilities on the other. For example, there is no regeneration in nematodes, whether small or large (their length ranges between 80 micrometres and 8 metres). Molluscs, on the other hand, have good regenerative capacities. Octopuses can completely regenerate a lost arm, even repeatedly. Oysters can regenerate gills and other parts of the body and also completely renew the entire shell; snails regenerate part of the head, including tentacles and eyes. Several species of annelids even regenerate gonads and germ cells.

Finally, regeneration is not limited to multicellular organisms. In a large ciliate such as *Stentor coeruleus*, a millimetre long, a small cell fragment can regenerate, provided it contains sufficient nuclear material. If we needed good reasons to include unicellular organisms in the subject index of developmental biology, here we have one.

Even if the product of regeneration is often a good copy of the missing part that is thus replaced, the process does not seem to depend on a memory of pre-existing structures. The most convincing examples are offered by animals that reconstitute themselves from the re-aggregation of previously isolated cells. In an experiment on hydra, cells of the tentacles or located near the mouth were marked before being dissociated, but in the animal reconstituted after dissociation and re-aggregation these cells were found dispersed in various parts of the body of the new polyp.

In this experiment, of course, hydras were subjected to conditions they would have never experienced in natural conditions. This objection applies to much of the experimental work on animal regeneration. Caution is required in interpreting the results, but we can learn an interesting lesson from that. It is not possible to invoke evolutionary explanations – in terms of adaptation – for an animal's response to these unnatural situations; even less meaningful would be to invoke the presence, in an animal's genome, of a specific programme for a response to conditions that the species has never experienced in its evolutionary history. It is more reasonable to describe these regenerative processes as 'inertial' responses of the living system, based on generic properties and not specific to regeneration.

What do we know about regeneration in terms of developmental genetics?

Very large numbers of genes are expressed in regenerating tissues at particularly low or particularly high levels. In the marine polychaete worm *Syllis gracilis*, complete anterior regeneration involves differential expression of about 2000 genes; in a related species, *Sphaerosyllis hystrix*, in which anterior regeneration is incomplete, 4771 genes are involved. There are no estimates of the total numbers of genes in these animals, but we can guess that about 20% of the total are involved in regeneration. This is in line with the result of a study on changes in gene expression associated with the regeneration of the intestine in the sea cucumber *Apostichopus japonicus*: here, increased expression was found for over 2400 genes, decreased expression for over 1000, the two numbers estimated to be respectively about 10% and 5% of the total.

Regenerative abilities are generally modest, or very modest, in vertebrates and in arthropods. In both groups, regeneration is essentially limited to the appendages, and very often it is also missing in these. Salamanders can regenerate limbs and tail, lizards only the latter. Even in arthropods, regeneration is essentially limited to the appendages and is practically absent in many groups.

In some flatworms, regeneration can prolong life. In an experiment on the tiny (1.7 millimetre) marine flatworm *Macrostomum lignano,* an individual was cut transversally 29 times and regenerated as many times, thus reaching an age of 12 months, 2 months more than it would have lived in normal conditions. In light of this result, we should reverse the popular, finalistic view that many animals regenerate to live longer, to say instead that some animals live longer because they are able to regenerate.

Closing the Circle

The Egg Is Not the Least Specialized of Cells

Three weeks after the start of incubation, the chick is ready to break through the eggshell and to begin active life. By interrupting incubation at different times, we could have seen, in the previous days, the progressive emergence of the embryo and its subsequent development into chick. At the time the egg was laid, however, the egg contained only dense fluids, with no evidence of the future animal.

For centuries since Aristotle's pioneering studies, the chicken's egg has been the most accessible object on which to study an animal's early development. It is understandable, therefore, that the very notion of development has long been based on observations made on this bird. To the eye of the observer, the chick takes form gradually. At the beginning, no sign seems to pre-exist of the complex structure that will emerge gradually. Moreover, the isolation of the embryo inside the shell excludes any shaping intervention from outside. Chick development therefore supports the theory of development by epigenesis. To understand the meaning of this expression (literally, genesis by additions) it should be said that the term 'epigenesis' is used here in the meaning that has been attributed to it for over 2000 years; it should not be confused with the epigenetics of recent decades (p. 54). The epigenetic theory of development is the alternative to pre-existence (p. 81).

Regardless of the conflicting interpretations of the nature of developmental phenomena, advances have reinforced the notion that the starting point for the development of a new individual is the egg. For a long time, uncertainty remained about mammals, which are viviparous and thus did not provide easy access to the earliest stages of their individual development. Do they also produce eggs? The answer to this question was provided in 1827 by Karl Ernst von Baer, who described the ovum of the dog. The egg of the human species would be described a century later. The fertilization of an egg by a spermatozoon was observed for the first time in 1875, by the German zoologist Oskar Hertwig.

Shortly after von Baer's discovery, the cellular theory of Schleiden and Schwann took hold (p. 43). At this point, it seemed natural to generalize about two points: first, the development of all animals starts from a particular cell, the egg; second, the egg is a sort of universal (generalized and perhaps ancestral) cell, devoid of all the specializations that characterize the different types of cells that make up an animal's body. A little later, with the emergence of the theory of evolution, the egg, the single cell from which development begins, would start to be compared to the single-celled ancestors of all multicellular animals.

These two generalizations, however, are not justified. The first neglects those animals that are formed from buds or other parts of the parent's body (p. 76); the second attributes to the egg prerogatives that do not belong to it. I address this rather serious problem in the following paragraphs.

First of all, eggs are not all the same. The most obvious differences between one egg and another are their sizes (from a fraction of a millimetre to the several centimetres of the eggs of ostrich and swan) and the presence or absence of a protective shell (rigid in birds, elastic in lizards, none in almost all aquatic or viviparous animals). We will see in a moment that there are also differences in the basic cellular processes (meiosis, or lack of it) that characterize oogenesis.

More important, however, is the fact that the egg is one of the most elaborate cell types, the result of a process that deserves to be described as the development and differentiation of a single-celled organism.

This is confirmed by the fact that a specialized set of genes are expressed during oogenesis. For example, in the oocytes of sea urchins, genes are expressed that are required for the growth of the egg, genetic recombination and meiotic division, storage of nutrients and fertilization. Clearly, this does not match the description of a hypothetical 'generic' metazoan cell. Throughout oogenesis, the egg cell and its immediate precursors have a rather passive role, strongly influenced by the mother. The origin of the yolk, responsible for the often unusually large size of the egg cell, is maternal; also of maternal origin are many of the mRNA and protein molecules that will control the first stages of development of the embryo, until the cells of the latter are able to express their genome (p. 83). From the moment a cell begins differentiating as an egg until the moment it divides into two daughter cells (as a rule, after fertilization), the egg goes through a much more extensive series of changes than almost any other cell.

Tradition aside, there is no reason to deny oogenesis the title of a developmental process. If necessary, another argument in favour of this interpretation is that some animals produce two different types of eggs, and through different processes. This happens in all plants and animals with facultative or alternating parthenogenesis. An example is offered by monogonont rotifers.

These are tiny animals, whose body is made up of about a thousand cells (roughly like *Caenorhabditis elegans*). Monogonont rotifers are common in ponds, lakes and other aquatic environments. Their biological cycle is quite complex. During the favourable season, the population includes only diploid females that reproduce by parthenogenesis, producing exclusively female offspring. These daughters derive from eggs that were not produced through meiosis and therefore possess the entire diploid chromosome number of their mothers. At the end of the favourable season, however, specific environmental signals induce females to produce, through regular meiosis, eggs with haploid chromosomal number. All individuals developing from these haploid unfertilized eggs are male, but within a short time these are ready to produce sperm that can fertilize haploid eggs similar to those from which they developed. Fertilized eggs will not develop until the adverse season is over, then they will eventually give rise to the first generation of parthenogenetic diploid females in the new year.

To conclude, if we are really looking for 'generic' cells, we will not find them in the eggs but, if anywhere, in the stem cells.

Non-Adaptive Development

Development Is Not Necessarily a Sustainably Adaptive Process

A lot of interesting things we know in evolutionary biology have been learned by studying island populations, such as Darwin's finches and giant tortoises of the Galapagos and the hundreds of Hawaiian species of *Drosophila*. Still, we know how uncertain survival is on an oceanic island, and how many species have disappeared from the Galapagos and Hawaii, not always because of human disturbance. It is possible to predict that all the species that have evolved on those remote archipelagos may become extinct in the future without leaving descendants. If this is the case, we will have to accept that precisely the populations that best lent themselves to the study of evolution are those that live in places that are not very suitable to ensure the future of their lineages.

Somewhat similarly, developmental biology continues to progress thanks to the study of organisms that have no hope of reaching adult stage, or will reach it in conditions that we can describe as desperate. The study of 'monsters' led to the discovery of the *Hox* genes, which have a key importance in placing the different organs in their position along the antero-posterior axis of the body. Mutations in some of these genes can result, for example, in the development of a pair of legs on the head of a *Drosophila*, in the place normally occupied by the antennae. Lacking these important sense organs, the little fly will have problems finding food or a partner. In other words, this mutant fly is an individual with no future, a failure in life, like a species that becomes extinct on a remote oceanic island from which it cannot escape.

A fruit fly without antennae is thus a lost individual, from the perspective of its population. However, it can be a source of important knowledge on matters of developmental biology.

These examples allow us to take up a topic introduced at the start of this book: the distinction between development as process, and development as the story of the individual that, in the most familiar situation, goes from the egg to the adult, with all the exceptions and distinctions that fill the pages of this book.

A caterpillar unable to turn into a chrysalis can be a choice subject on which to study a very important stage in the development of insects – their metamorphosis. The same caterpillar is an example to be discarded, if we are interested in the history of individual development. It is a failure, indeed, in contrast to the regularity of the changes that accompany development from egg to adult, all of which are expected to be adaptive.

It is not easy to accept the existence of maladaptive developmental processes. But this is a fact with which we must cope, including the most worrying case, cancer. The progression of a tumour is a developmental process, even if, for the victim, its adaptive value is negative.

We can also say that the progression of a tumour is, at least in the short term, adaptive for the cells that make up the tumour mass. If anyone remains unconvinced that the study of cancer belongs to developmental biology, perhaps they will change their mind after reading the following story.

The Tasmanian devil is a robust Australian marsupial with a rather impressive appearance (almost completely black, with a large mouth armed with 42 teeth, and carnivorous habits), whose populations have undergone a steep decline in the past 25 years owing to a devastating tumour that attacks its face, making feeding impossible. This cancer is transmitted between Tasmanian devils when they bite infected individuals of their species. As a consequence, this line of cancerous cells does not go extinct with the death of the individual on which it occurs, but propagates indefinitely, behaving like an autonomous biological species. For this cell line, the development is certainly adaptive. Transmissible tumoural cell lines have also been found in dogs and in bivalve molluscs.

If we focus on the cellular mechanisms involved in these processes, it is difficult, perhaps impossible, to determine a precise divide between adaptive developmental processes and pathological morphogenetic events such as carcinogenesis. In addition to these, some researchers believe that the inflammatory events typically described in immunological or pathological contexts are also relevant to developmental biology.

Back to individual development, we must also recognize that development under standard conditions is not necessarily optimal in all respects. For example, the legume *Anthyllis cytisoides*, a small shrub from southern Europe, grows better on pastureland than in undisturbed areas. More convincing is perhaps the case of regeneration, which sometimes imposes heavy compromises. In sponges and corals, regeneration has negative effects, often strongly so, on "normal" processes such as somatic growth, sexual reproduction and the ability of these animals to defend themselves from predators, to face competition and to recognize conspecifics.

In the latter examples, the suboptimal character of the developmental process can be explained in terms of the conflict between cell dynamics and contrasting metabolic needs.

6 Genes and Development

Developmental Genes?

In the Genome There Are No Specific Genes for Specific Phenotypic Traits

In 1994, the science journal *Nature* hosted a Scientific Correspondence page entitled "Inversion of dorso-ventral axis?", written by two German researchers, Detlev Arendt and Katharina Nübler-Jung. The meaning of the title might have been clear to biology history scholars, for reasons we will see immediately, but hardly so to developmental biologists. However, the short article eventually opened new horizons in developmental genetics. This chapter is entirely dedicated to this discipline, and it is fitting to start by summarizing the first lines of the article by Arendt and Nübler-Jung. They proposed a revisitation of a long-established point of comparative anatomy and evolution, according to which the main axis of the nervous system in insects and vertebrates, which is ventral in the former but dorsal in the latter, is regarded as the product of independent evolution. Arendt and Nübler-Jung proposed a different interpretation: the main neural axis of both kinds of animals would derive from the same centralized nervous system in their common ancestor. If so, the ventral side of insects would correspond to the dorsal side of vertebrates. This idea was not new: it was first floated by Étienne Geoffroy Saint-Hilaire, the father of the Isidore Geoffroy Saint-Hilaire cited in the first pages of this book.

In 1822, Geoffroy Saint-Hilaire the elder proposed this equivalence between arthropods and vertebrates as an example of his theory that all animals are built according to a common body plan. This means, for example, that the relative positions of the gut and nervous system are the same in insects and vertebrates. The idea was too bold for the time and, it must be said, purely speculative. Now, however, Arendt and Nübler-Jung could support it with an important piece of developmental genetics: the *decapentaplegic* (*Dpp*) gene, which is expressed on the dorsal side in the *Drosophila* embryo, and the *Bone-morphogenetic-protein-4* (*BMP-4*) gene, which is expressed on the ventral side of the vertebrate embryo, are homologous: in less precise terms, we could say that they are two versions of the same gene.

The short 1994 article immediately attracted general attention. At stake was not only the late vindication of Geoffroy's theory, but also the probable need for a radical revision of current ideas regarding the role of genes in developmental processes. If two animals as different as an insect and a vertebrate have homologous genes controlling dorso-ventral polarity, one should perhaps think that the role of genes in developmental processes is much less specific than previously thought. We will return to this shortly.

Meanwhile, let's mention that the idea of an equivalence between the ventral side of vertebrates and the dorsal side of insects and other invertebrates (and vice versa) was soon confirmed and better specified, in particular in a 1996 article published by the American embryologist Edward M. De Robertis who, together with a young Japanese collaborator, Yoshiki Sasai, demonstrated that in addition to *Dpp/Bmp-4,* the dorso-ventral polarity is controlled by the expression of another gene: this is known as *short gastrulation* in *Drosophila,* where it is expressed on the ventral side of the embryo, and as *Chordin* in vertebrates, where it is expressed on the dorsal side of the embryo.

It is not surprising that the expression of the same gene (more precisely, of its different homologous variants present in different animals) can have opposite effects, such as these dorsalizing and ventralizing genes. In plants too, genes are known whose expression is 'interpreted' in the opposite direction in different species. For example, in the buds of many flowering plants, the expression of the different homologous variants of the gene YABBY mark the outer face of the leaf in the bud (the side that will give rise to the lower face of the leaf), but in maize it is the opposite face that is similarly marked.

These examples demonstrate the distance between contemporary genetics and the science of heredity in Mendel's time. It is not simply that today we know the molecular nature of those hereditary factors, already postulated by

Mendel, which in 1909 received the name of genes by Wilhelm Johannsen (cf. p. 54). What is more important is that, while for a long time it was possible to study only the way in which genes are transmitted across generations, today we can also study the way in which genes are expressed and the effects of their expression on development.

At this point, we can ask if there are real 'developmental genes'. The answer is yes, but a qualified yes. In fact, the expression patterns of many genes are strictly limited and correlated with specific times, places and events in development. Mutations in these genes can alter the normal course of development, sometimes with dramatic effects on the resulting phenotype. However, we must refrain from taking any of them as 'the gene for' the shape of a limb or the symmetry of a flower. Furthermore, we need to consider any particular gene in the context, not only of other genes, but of the entire cellular (and multicellular) environment.

A Genetic Programme?

Development Is Not the Actualization of a Genetic Programme

In an effort to learn about developmental processes and to explain their mechanisms, researchers have progressively used the different and increasingly powerful tools that technology could offer: from the simplest microscopes of the seventeenth century to the different types of electron microscope of our day, from simple devices for mechanical manipulation of embryos to the sophisticated equipment of a modern developmental genetics laboratory, where it is possible to study the effects of the expression of single genes and networks of genes.

Technology has always offered something else, beyond the tools of the daily lab work, and that is inspiration for theories of living beings: how these can work, how they can be built.

For centuries, the mechanical model capable of inspiring these interpretative efforts has been the watch: a sophisticated machine that only a few particularly skilled craftsmen were able to build. Many automata have also come out of the hands of some of these craftsmen, automata often disguised behind a human-like figure and capable of reducing, in the imagination of many, the distance between the machine and the living – indeed, even between machine and human.

In today's automata, however, there are no cogwheels and other gears of mechanical watchmaking, but computers equipped with programmes and capable of responding to a variety of information that the sensors of an automaton (today we prefer to call it a robot) collect from the external environment. If the features and behaviour of a human being are imitated much better in modern robots than in the mechanical automata of the past, computer-based technology has suggested, in return, new models for a theory of living and development.

In 1994, Lewis Wolpert dared to say that perhaps, if we work hard, we will be able to 'calculate' an embryo. More precisely, his question was whether, starting with a comprehensive description of a zygote (its genome and the identity and distribution of all protein and RNA molecules), one could predict the future development of the embryo. Wolpert was well aware of the difficulty of the task. In principle, it would require the behaviour of all the constituent cells to be computed. Therefore, he suggested a less complete but more realistic version of the project, by choosing to describe cell behaviour at the lowest level of complexity adequate to account for development.

The idea that an embryo can be 'computed' may seem alien, but it is a direct consequence of the popular notion that a programme to build a living being (and also, to a certain extent, to make it work) is contained in its DNA. The expression 'genetic programme' came into use in 1960; the following year, the molecular biologists François Jacob and Jacques Monod described the passage of genetic information from DNA to mRNA as 'transcription'; in 1963, the metaphor was completed by naming as 'translation' the assembly of the amino acids that make up a protein according to the order 'dictated' by mRNA. Since then, the genetic programme metaphor has been increasingly successful, up to its current use (and abuse) in common parlance.

Before the end of the last century, however, signs of disagreement began to be perceived around the use of the metaphor of the genetic programme.

A fundamental objection was that DNA alone does absolutely nothing. Viruses demonstrate this (it must be said that in many viruses the so-called genetic information is encoded in an RNA molecule, not in DNA; but this does not matter here): a virus does nothing until it contacts a host cell, whose transcription and translation machineries are necessary for the expression of the viral genome. Without rejecting the notion of a programme, but reversing the possible roles of nucleus and cytoplasm, Evelyn Keller suggested in 2000 that genetic information could be compared not to a computer programme but to a set of data that will be processed according to a programme present in the cell structure. This programme may reside in the transcription and translation mechanism. In this objection, however, we remain within the scope of computer-inspired models, whereas it is better to go back to the facts of biology.

The genome should not be understood as a simple catalogue of genes to which correspond as many different mRNA molecules and as many different proteins. Thanks to a mechanism known as alternative splicing, many different forms of mRNA and therefore of proteins can be obtained from the same gene: for the *Drosophila* gene *DSCAM* these are estimated to number about 38 000. This is possible because of the way the mRNA transcribed from these genes is processed before being translated into protein. I mentioned earlier (p. 24) the huge amount of time required for the transcription of very large genes. Their length is many times bigger than the segment containing the information corresponding to the protein that will eventually result from the translation process. The whole gene, and the mRNA as transcribed from it, contain traits (exons) the sum of which represents the actual gene (that is, the sequence of nucleotides that encode a protein), and traits (introns), often very long, that will be cut away during a process called the maturation of the mRNA; only the edited molecule will be translated.

At any rate, the most serious objection to the genetic programme metaphor is the absence of a simple and unambiguous relationship between genotype and phenotype. It is not even possible to establish a clear dividing line between developmental processes controlled by genes and those that are modulated by environmental factors. I will discuss this in Chapter 8 (p. 139).

If we still need proof that in the genome there cannot be a complete and accurate description of an organism, a glance at neural networks (those in our brain, for example) or at blood vessel systems is sufficient. In Chapter 7 (p. 122) we will see how nerve fibres find their way; here I will devote a few words to the circulatory system. There is no doubt that in the construction of the complex network of vertebrate blood vessels some aspects are strictly controlled by genes: for example, whether a vessel will be a vein or an artery, or how to make branches and anastomoses (connections between branches). However, the actual course of the vessels, including the determination of the precise points of branching and anastomosing, are not genetically fixed. After all, our genome is too small to contain such detailed instructions. This degree of freedom in the creation of the vascular network explains to a large extent the differences we observe, not only between individuals, but also between the left and right halves of our body: a clear example can be seen by comparing the superficial vessels visible on the inner face of our wrists.

Some researchers have shown the important role of blood pressure in shaping the vascular network. In the first week of development of the zebrafish embryo (I mentioned this fish in Chapter 1, p. 15, among the fashionable model organisms), in the midbrain a very branched vascular network is formed, which then undergoes robust pruning controlled mainly by the blood flow and operated by extensive migration of endothelial cells (those that form the inner lining of blood vessels) rather than by cell death.

Gene Networks and Master Control Genes

Master Control Genes Responsible for the Production of Major Organic Structures Such as Eyes and Hearts Probably Do Not Exist

In the early years of the last century, with the launch of the science of heredity following the rediscovery of Mendel's work, nobody thought that the results published in 1866 by the Bohemian friar would be valid only for peas, or for plants generally. It was much more reasonable to think that in all kinds of organisms there were material entities responsible for the transmission of characters from one generation to the next (the genes, as these entities were called later; see p. 96). The molecular nature of the gene was demonstrated in 1952 by Alfred Hershley and Martha Chase, with experiments performed on the T2 phage, a virus parasitic of bacteria. In the same year, the discovery of the three-dimensional structure of DNA was published. Six years later, Francis Crick would formulate the so-called central dogma of molecular biology, the principle according to which genetic information can only pass from DNA to mRNA and from the latter to proteins, but not the other way around: a principle of the highest interest, although the use of the term dogma was inappropriate, in a scientific context, and despite the exceptions to its alleged universality subsequently discovered.

However, this formidable progress ignored the comparative aspect of biological phenomena. That is, it was not possible to answer questions like these: if in humans there are genes that control the number of fingers or toes, are these the same genes in all terrestrial vertebrates, including those with a different number of fingers?

In fact, it would not have been possible to address such questions until it was at last possible to study the ways and times that these genes, whose molecular structure was beginning to be known, were expressed during development. When this target became accessible, a most sensational surprise was that many of these genes are the same in organisms as different as the mouse and the fruit fly.

This expression, 'the same genes', must be taken with flexibility. Even within one species, each gene is present, with negligible exceptions, in forms (alleles) that differ to varying degrees which can result in differences between the individuals, but more often do not involve any difference in the phenotype. On the other hand, all living species derive from a common ancestor, so there is no reason to suppose that each species has its own genes. Rather, since the differences between copies of the same gene tend to accumulate over time (if they are not wiped out by chance or natural selection), we must expect that, on average, the differences between the copies of the same gene present in two different species are larger the more distant in time is their last common ancestor. For example, the genes of humans will be more similar to those of the chimpanzees than the genes of either chimpanzees or humans are similar to those of cats or sea urchins.

It would not be surprising, therefore, to find that the same character is controlled by the same gene in humans and chimpanzees, but from studies of about 40 years ago there emerged the identity of genes involved in important stages of development in animals as diverse as mice and *Drosophila*.

The discovery that eye morphogenesis is controlled in part by the same genes in animals so different and so distantly related as squids, insects and vertebrates made particular outcry. One of these genes is known in vertebrates as *Pax6*, while its equivalent in *Drosophila* is best known as the *eyeless* (*ey*) gene. The eyes of vertebrates and insects are built according to very different structural and functional principles, so researchers of comparative anatomy had never imagined that they might all derive from the visual sense organ of a remote common ancestor. This idea, however, has been supported by some developmental geneticists, precisely on the basis of the universal involvement of the *Pax6/ey* genes in eye morphogenesis.

However, the expression of *Pax-6* is not limited to the eyes but extends to a large part of the nervous system and sense organs. This applies to both vertebrates and *Drosophila*. Even more critical is the presence of *Pax6* homologues in animals that do not have eyes, such as sea urchins and nematodes, including, of course, *Caenorhabditis elegans*. Some authors believe that this gene, originally expressed in the front part of the body, has been co-opted by different evolutionary lineages independently, in the construction of the eye.

This gene, however, does not act alone: in *Drosophila*, for example, it has been estimated that the expression of about 2500 genes is involved in the construction and maintenance of the eye, equal to 18% of the total number of genes in this insect.

Indeed, a large number of genes are expressed in all body parts. In animals, about half of the genes are probably expressed in the nervous system. Many genes encode proteins that function as regulators of the expression of other genes. The genes that code for these proteins, known as transcription factors, are estimated to be around 12% of the total number of genes in *Arabidopsis thaliana*, and 18% in baker's yeast.

Undoubtedly, the effect of the expression of some genes is more important than the effect of the expression of others. At the beginning of the 1990s, the geneticist Edward B. Lewis introduced the concept of the master control gene, to indicate the key genes, the expression of which is essential to start the expression of the entire constellation of genes that characterizes an important morphogenetic event. The champion of this notion, however, was the Swiss developmental geneticist Walter Gehring. In favour of this notion, Gehring adduced not only the data of the comparative genetics of development (with *Pax6/ey*, of course, regularly presented as the typical master control gene), but also the remarkable results of experiments carried out by his team in Basel. With complex manipulations, these researchers managed to express the *ey* gene in a body part where, in nature, it is never expressed – on the legs of

Drosophila. The result was sensational: the fruit flies thus manipulated developed on their legs accurate copies of a compound eye, even if reduced in the number of facets. As often happens, however, research progress in this field led in a few years to the *eyeless* gene being dislodged from the alleged role of master control gene for the production of the eye in *Drosophila*. Six other candidates (*twin of eyeless, eyes absent, sine oculis, dachshund, eye gone* and *optix*) were proposed in its place, but in the end the notion of master control gene was abandoned by many. Among these was the American developmental biologist Eric Davidson, who in the last years of the past century demonstrated how morphogenetic processes are controlled by complex networks of gene interactions, mediated by the transcription factors produced by many of them, rather than by a linear control sequence that would begin with a hypothetical master gene.

Gene Expression throughout Development

Genes Expression Relevant to Development is Not Limited to the Embryonic Phase

The discovery of genes common to the most diverse animals, in which their expression is equally involved in the control of important morphogenetic events, is not limited to *Pax6* and its role in the production of eyes. The role of the *tinman* gene in relation to the heart is similar. The most extensive and also most unexpected similarities between the genes that control important aspects of morphogenesis are, however, those concerning the *Hox* genes, discovered and characterized since the early 1980s as the genes whose expression identifies positions along the main body axis of almost all animals. I will return to this in Chapter 7 (p. 117).

For some time, it was believed that their expression was confined to a narrow time slice of early development. However, this belief was supported only in part by experimental evidence. Far more important was the fact that expression of these genes at later developmental times was not documented simply because it had not yet been searched for.

In more recent times, the attention of developmental genetics scholars has widened somewhat. Thus it has been seen that expression of the *Hox* genes and other important genes in morphogenesis is not restricted to the embryonic phase of an individual's history.

For example, in the abalone *Haliotis asinina*, *Hox* genes homologous to those that in *Drosophila* specify positions in the anterior section of the main body axis (genes known in *Drosophila* as *labial*, *proboscipedia*, *zen*, *Deformed* and *Sex combs reduced*) are expressed in the mollusc larva. The expression of two of them (the homologues of *labial* and *Deformed*) is limited to the mantle margin, where are located the cells that secrete the material from which the shell is made.

Studies of this type, however, do not tell the whole story of individual development from the point of view of gene expression. In the first years of this century, systematic screenings of gene expression have at last been carried out at many times between the beginning of embryonic development and adulthood in a model species. The first detailed and informative study was, as expected, on *Drosophila melanogaster*: of all the best-known model animals, the fruit fly is the species that undergoes the most radical metamorphosis and is thus particularly informative.

The embryonic period is the one in which changes are most rapid, so the study was performed on embryos sampled at 1-hour intervals; sampling was less closely spaced for the larval and pupal stages. Adults were also examined, sampled at 1-day intervals. In all, 66 steps of the fruit fly's developmental sequence were compared.

In this study, the level of expression (in the form of mRNAs resulting from gene transcription) was examined for 4028 genes, more than a quarter of the total number of genes estimated to be present in the insect's genome. Of these genes, 86% showed significant changes in the level of expression during development; in almost all cases, the highest level of expression was at least four times higher than the lowest level. Of these variable expression genes, 88% are expressed in the first 20 hours – that is, during embryonic development.

However, most genes that have a high transcription peak within the first 2.5 hours from the start of embryonic development have a second peak in the transition phase between the last larval stage and the pupa. On the other hand, the pattern of gene expression observed in the larval phase is not very similar to that of the pupa, but it resembles that of the adult. Finally, important variations were observed also within each phase. As expected, the pattern of gene expression during embryogenesis (with important changes at the beginning, in the middle and at the end of embryonic development) and during the pupal phase is particularly dynamic, even if these moments are shorter than the larval and adult stages.

Only a fraction of all these genes have a precise identity – that is, a name and, more important, a well-characterized function. Only a small percentage of them fall into the category of recognized developmental genes. Nevertheless, just from the results of this study and other similar ones, performed in recent years on different animals, there are some indications that lead to the revision of traditional interpretations. First, an important fraction of the entire genome is expressed at all times along development and in every part of the body. Second, the level of expression of individual genes varies in a characteristic and more significant way in the moments in which the morphogenetic changes are more rapid and intense, such as the embryonic phase and metamorphosis. Third, a dynamic expression of an important fraction of the genome is by no means restricted to these phases and could be involved in stabilizing the phenotype.

This last suggestion throws light on one more deficiency in the traditional notion of development: the absence of obvious morphological changes (a condition we may call morphostasis) is not necessarily a sign of suspension of development; it could instead express a condition of balance between opposite dynamics. Here, too, a comparison with mechanics is useful: a state of rest is not necessarily uninteresting when it is not the system's inertial condition, but the effect of constraints or of balance of contrasting forces.

Painting Modules Differently

The Whole Organism May Not Develop as a Tightly Integrated System

Among the authors who contributed most to the transition from traditional embryology to modern developmental biology was the British biologist Conrad Hal Waddington. Among other things, he was responsible for identifying a process (genetic assimilation) that allows some acquired traits to become hereditary through a mechanism that does not conflict with evolution by natural selection. More interesting in the context of developmental biology is his concept of canalization, the ability of an organism to produce the same phenotype despite variations in genotype or environment.

Waddington's epigenetic landscape is a metaphor for how gene regulation modulates development.

Waddington was the author of books such as *Principles of Embryology* (1956) that have left a deep trace in the history of biology. Among the developmental processes covered in that book, there is what Waddington called regionalization, which he defined as the emergence of different parts within an originally uniform expanse of tissue. In the most recent scientific literature, this is usually described as pattern formation rather than regionalization, but the phenomenon is still the focus of a lot of research. The emergence of the differences between cervical, thoracic and lumbar vertebrae in our spine is an example of pattern formation, as is the precise positioning of seven black spots on the orange background of the elytra of the most popularly known (in Europe) of the many species of ladybirds.

Another term widely used in our days is 'module', but – as often happens in biology – there is no precise agreement on its use. In general terms, a module is a part of a larger system (in our case, of a living organism) made up of subordinate units (for example, cells) that interact with each other or are involved in a given process much more strictly than they are with parts of other modules. The integration between the parts of a module can be observed at different times and concern different phenomena. From a morphogenetic point of view, each bone of our skeleton can be considered a module. At a different level, the whole skeleton can be described as a module distinct, for example, from the circulatory or muscular apparatus, but from a functional point of view it is more useful to consider modules such as the hand, the arm or an entire limb (or a pair of limbs).

The boundaries between two modules that are differentiating in different ways are often virtual, like the borders between groups of people who, at a popular party, are talking to each other without paying attention to people in the other groups.

At other times, however, boundaries take on physical consistency. This is observed, for example, in the vegetative apex of plants. Here, between the cells from which a young leaf is taking shape and the stem cells in the bud that sprouts at the axil of the leaf (a future flower or branch), there are thin layers of cells of particular shape and with reduced growth that act as a border.

In the embryonic development of many animals, an early sign of modularization is the split between somatic and germ line cells. The latter will give rise to the gametes; therefore, they are the only cells that may have descendants in future generations, while the progeny of the somatic cells – whose indirect contribution to the reproductive success of the gametes may be nevertheless indispensable – will die out with the death of the individual.

In some animals, the specification of primordial germ cells (PGC) is the result of the accumulation, in these, of gene products supplied by the mother. In *Drosophila*, for example, this material forms the so-called germ plasm, which occupies a part of the egg corresponding to the future posterior part of the embryo.

This situation corresponds quite well to the idea first expressed in 1892 by the German biologist August Weismann, to whom we owe the notion of a primary distinction between germ line and somatic line. However, Weismann believed that the germ plasm, rather than being specified by the mother in each generation, was a special immortal substance transmitted via germ cells from generation to generation. In any case, during egg cleavage, the germ plasm accumulated in the oocyte is sequestered inside cells that become, in fact, the PGC.

In the tiny wasps whose prodigious polyembryony I mentioned in Chapter 3 (p. 43), a germ line is recognizable after the second division: of the four blastomeres thus formed, one contains germ plasm (known to contain transcripts of the *vasa* gene): this is the insect's first PGC. This cell proliferates, as do the remaining (somatic) cells, until the time that the primary embryo fragments into more than 2000 secondary embryos, most of which contain at least one PGC; those in which no PGC is found develop into sterile larvae that do not complete metamorphosis.

In mammals and in many other animals, the mother has no role in the specification of PGCs. In this case, there is no germ plasm in the egg, and the PGCs will be specified later: in the mouse, for example, this happens around the beginning of gastrulation.

Before leaving this topic, we should mention that in plants there is no distinction between somatic line and germ line.

Back to animal embryos, the first global distinction that occurs among the somatic cells is usually their differentiation into two or three macromodules, the germ layers. In this case too, as in the germ/soma distinction, the divergence of fate between the cells belonging to the different germ layers is anticipated by the presence in each of them of specific molecular markers, the products of different genes. Mesoderm cells, for example, are characterized by the expression of two genes called *snail* and *twist*.

In most animals, each germ layer gives rise to the same distinct set of organs and tissues, as mentioned in Chapter 5 (p. 78). However, this is not a universal rule. In hydrozoans, which have only two germ layers (ectoderm and endoderm), germ cells generally differentiate from ectodermal cells, but sometimes from both layers. In the hydrozoan *Clytia gregaria*, the nerve cells – which in almost all animals are of ectodermal origin – derive instead from the endoderm. In most metazoans, the midgut derives from the endoderm, but in tardigrades (tiny invertebrates also known as water bears) it is of mesodermal origin. In many crustaceans, including woodlice, the ectodermal cells of the posterior intestine proliferate by expanding in the anterior direction and replace the original endodermal primordium that eventually disappears.

That's not all. In planarians, during embryonic development there is no clear evidence of germ layers or organ primordia. From this point of view, a planarian embryo resembles the blastema from which regeneration starts. As shown by an exemplary study published in 2005 by Albert Cardona and colleagues, in planarians there are indeed many similarities between embryonic development and regeneration of lost body parts. In both instances a provisional epidermis is first formed, followed by multiplication of stem cells, out of which the definitive epidermal cells differentiate which progressively replace the provisional ones, intercalating between them. Even closer similarities between the two processes are seen in the formation of muscle tissue and in the processes by which the entire architecture of the nervous system is set up.

In many zoological groups, after the separation of germ and soma and the identification of germ layers but still in quite early stages of development, groups of cells are singled out that will have a fate so peculiar as to be, in a sense, in competition with the remaining cells of the animal. In the larvae of these animals, there are groups of cells that do not take an active part, for the moment, in the formation of the animal's organs. These set-aside cells will proliferate at a later stage and give rise to the adult's organs, while the larval structures are dismantled. These animals undergo metamorphosis that drastically refreshes the body's architecture.

Set-aside cells are found in many marine animals, for example in sea urchins and sea stars, and also in the ribbon worms mentioned in Chapter 4 (p. 70): the reader will remember that in some of them an adult worm originates from the coalescence of eight separate groups of set-aside cells.

The cells that form the imaginal discs of holometabolous insects – those that undergo a 'complete' metamorphosis with a resting stage (the pupa) between larva and adult – are a kind of set-aside cells. Examples are flies, beetles, butterflies and bees. Imaginal discs are flat bilayers of cells that proliferate slowly during larval life and eventually give rise to the characteristic organs of the adult, such as wings and genital structures. How much of the larva's body will be destroyed at metamorphosis, and what will instead be constructed from imaginal discs and other cell groups known as the histoblasts, varies greatly in the different groups of holometabolous insects. The most radical transformations are observed in flies.

In *Drosophila*, the cells that will form the imaginal discs are set aside very early in development. After the last embryonic wave of mitosis, larval cells become polyploid (that is, their nuclei contain multiple copies of the whole set of chromosomes, beyond the ordinary diploid condition with two sets), cease to divide and differentiate into the different tissues and organs of the larva, whereas imaginal cells remain diploid, continue to proliferate and do not differentiate until metamorphosis.

The Developing Individual as a Cloud

One Organism Does Not Necessarily Contain Only One Genome

Popularly known as the bobtail squid, *Euprymna scolopes* is a small cephalopod mollusc, at most 8 centimetres long, belonging to the sepiid group and thus more properly a cuttlefish rather than a squid. This species has neither the economic importance of other cephalopods, nor the huge size that often puts giant squid in the headlines: instead, it has become known in recent years for its singular symbiosis with a luminescent bacterium. Many marine animals, especially among those living in dark deep waters, are capable of emitting light. The animal itself is often responsible for the chemical process from which energy is released as light; but in some cases, light is produced instead by bacteria hosted by the animal within specialized organs.

The production of light by the bobtail squid depends on symbiosis with the bacterium *Aliivibrio fischeri*. The squid has a specialized light organ, which is colonized by bacteria as soon as the mollusc leaves the egg envelope and begins active life. Cilia-bearing cells in the light organ attract bacteria, but allow only the cells of *A. fischeri* to settle there, while bacteria of different species are removed. A series of reciprocal interactions between squid cells and bacteria thus begins. Bacteria induce changes in shape and gene expression in the mollusc cells, thus actively contributing to the morphogenesis of the light organ.

This case, particularly well studied, is not the only example of the influence of symbiotic bacteria on animal morphogenesis. On the contrary, it is very likely that similar phenomena occur in the normal development of many animals, including humans. Furthermore, for many marine animals such as sponges, molluscs and annelids, the presence of bacteria on the substratum with which a larva makes contact can be the decisive factor that triggers metamorphosis (p. 11).

The mention of humans deserves a couple of words. Under normal conditions, our body is 'inhabited' by an incredible number of microorganisms, most of which belong to two kinds of bacteria (Bacteroidetes and Firmicutes). It is not easy to estimate their number accurately, but the cells that make up this microbiome are at least as numerous as the (human) cells of all our tissues. One recent estimate of the average number of cells in an adult individual of our species is around 37 trillion. To these should be added perhaps up to 100 trillion bacterial cells, the largest fraction of which lives in our digestive tract, followed by the population that settles on the skin. The composition of this microbiome varies from individual to individual, and it changes in relation to diet and health conditions and the drugs that may be used to keep these good. At any rate, it can be estimated that in our digestive tract there can be a thousand different bacterial species: the sum of their genomes gives a total number of genes equal to a few dozen times the genome of a human cell.

The human case is far from exceptional. In corals, more than half of the nourishment derives from tiny symbiotic algae of the genus *Symbiodinium*.

In ruminants, including domestic cattle, digestion depends on the activity of a heterogeneous and abundant set of microorganisms (bacteria, fungi and protozoans) that inhabit the animal's complex stomach and allow it to exploit cellulose as food. Cellulose is very abundant in plants, but the mammal's enzymes cannot digest it, whereas many bacteria can.

The microbiome can also have a specific morphogenetic role, and this is of particular interest for our discussion on the nature of developmental processes. We still know little about that, but available examples are very interesting. In small parasitic wasps of the genus *Asobara*, the ovary cells undergo apoptosis if they do not receive a particular chemical signal from symbiotic bacteria (*Wolbachia*). Also from *Wolbachia* comes the information necessary to correctly establish the antero-posterior polarity in the embryo of a tiny worm, the nematode *Brugia malayi*.

In an excellent 2012 article entitled *A Symbiotic View of Life: We Have Never Been Individuals*, Scott F. Gilbert, Jan Sapp and Alfred I. Tauber highlighted the universality of these associations between animals (and plants) and microorganisms: there is much more to symbioses than just lichens (p. 50).

The boundaries between host and symbiont are not always clear and, more important, not immutable. The mitochondria present in (almost) all eukaryotic cells and the chloroplasts characteristic of plant cells are the current versions of two different forms of bacteria incorporated at a very remote date by what became the eukaryotic cell. Over time, the interactions between the host cell and the hosted bacteria were not limited to changes in the structure of what gradually became two organelles, but also affected the genomes associated with them. Mitochondria and chloroplasts have their own genomes, residues of what was the genome of the bacterial chromosome, but a fraction of the genes that once belonged to this is no longer there, having been transferred to the chromosomes in the nucleus of the host cell.

Events of this type continue to happen. The cells of insects and other invertebrates host bacteria of the genus *Wolbachia*, in a relationship midway between mutualistic symbiosis (beneficial to both partners) and parasitism by the bacterium. In general, *Wolbachia* cells are easily recognizable within the host cell and appear to have preserved their entire genome; however, *Wolbachia* genes were found in the genomes of the *Aedes aegypti* mosquito and the seed beetle *Callosobruchus chinensis*, the result of recent transfers.

Genes from another symbiotic bacterium (*Buchnera*) have been found in the nucleus of aphid cells. Known cases are probably only the tip of a giant iceberg. Gene transfers can also occur between two organisms whose relationship is much less intimate than that of ancient endosymbionts with their host cells. In the case of the green sea slug *Elysia chlorotica,* which feeds on *Vaucheria litorea,* retaining the alga's functioning chloroplasts for months (p. 11), the relationship between the two organisms starts anew from scratch each generation; nevertheless, *Vaucheria* genes have been found in the mollusc's nuclear genome.

7 Emerging Form

Deceptive Numbers and Inelegant Morphogenesis

Morphogenetic Processes Are Not Always 'Elegant'

The first printed edition (1817) of Leonardo da Vinci's *Treatise on Painting*, published three centuries after the death of the great Renaissance man, contained only a part of the numerous fragments that Leonardo had entrusted to sheets of notes, currently split among a number of libraries. In particular, an entire section on botany for painters was missing, but it was eventually recovered and published in 1888 (in English) by the German art historian Jean-Paul Richter. Let's read some lines, the first ones from fragment 415, the next from fragment 417:

> Nature has so placed the leaves of the latest shoots of many plants that the sixth leaf is always above the first, and so on in succession, if the rule is not [accidentally] interfered with
>
> You will see in the lower branches of the elder, which puts forth leaves two and two placed crosswise [at right angles] one above another, that if the stem rises straight up towards the sky this order never fails

Only preceded by less precise accounts by Theophrastus and Pliny in classical antiquity, Leonardo's lucid account is the oldest accurate description of phyllotaxis, the spatial arrangement of the leaves around the stem. We can also speak of flower phyllotaxis, to indicate the arrangement of sepals, petals and stamens with respect to the axis of the flower.

As is clear from Leonardo's words, this arrangement follows precise geometric rules (Figure 7.1). These are not identical for all plants, but can be traced back to two main models. In the first, each leaf is attached to the stem at a different

(a)

Figure 7.1 (a) A plant (*Ajuga reptans*) with leaves inserted two per node, opposite to one another; those of the next node are inserted perpendicularly to the previous pair (decussate phyllotaxis). (b) A plant with spiral phyllotaxis (*Lithospermum purpureocaeruleum*). Leaves are inserted individually. In this example, the slightly bent leaf on the left is the fifth produced by the plant after the lowest leaf in the figure: the two leaves are found on the same vertical row, as in Leonardo's example.

(b)

Figure 7.1 *(cont.)*

point from the one that precedes it; in the second, two or more leaves are attached to each node. Leonardo specifically describes one common case of each of these phyllotactic models.

A whole range of models have been proposed to explain these regularities in the structure of plants, inspired by physics, by cell biology or, more abstractly, by number theory. The core of all these interpretations remains the so-called Hofmeister rule, named after the great German botanist Wilhelm Hofmeister, who proposed it in his work *Allgemeine Morphologie der Gewächse* (General Plant Morphology), published in 1868. According to this principle, each new leaf tends to form as far away as possible from the leaves developed (or developing) before it.

Be careful, though. We should not be dazzled by these regularities. The structural grain of natural objects, biological ones in particular, does not allow the precision of geometric figures. Moreover, we must also deal with plants that do not fit into these schemes, or switch from one phyllotactic pattern to another during their development. One of them is the 'model species' *Arabidopsis thaliana*: the two embryonic leaves (the cotyledons) and the two first vegetative leaves of this plant form two pairs of opposite elements (as in the entire elderberry plant, *Sambucus*, described by Leonardo), but the following leaves and even the elements of the flower (which to the naked eye seem to form regular whorls) are arranged individually along the axis, around which they describe a spiral, as in Leonardo's first example.

Geometrical regularities are also found in animals, behind which there may be more complex developmental processes than a mathematician would suggest as being necessary.

Let us start with animals whose body is divided into segments. The most obvious examples are perhaps earthworms and centipedes, but, beyond the external aspect, it is not difficult to recognize segmented organization even in vertebrates – it is sufficient to look at the spine.

It may be easy to produce a series of segments, even a long one, provided that a proliferative zone is available that can generate these structural blocks. The zone operates for a shorter or longer time. In this way, vertebrate somites are produced – the units corresponding to a vertebra and associated nerves, vessels and muscular units – but this does not necessarily apply to all segmented animals.

With increasing knowledge of the genetic patterns of expression involved in the production of geometrically elegant patterns, it is often discovered that these are not produced by the simplest mechanisms. For example, the earliest evidence of segmental organization in a *Drosophila* embryo lies in the seven stripes of expression of the so-called primary pair-rule genes. These stripes are produced synchronously, but each stripe is under separate regulatory control. This means that the *Drosophila* embryo is indeed *Making stripes inelegantly* – hence the title of an insightful paper by the British zoologist and developmental geneticist Michael Akam.

The number of segments that make up the body of leeches suggests all too easily an 'elegant' explanation, in the sense that mathematicians give to this adjective. These segments are always 32 in number, even if a first glance might suggest much higher numbers that differ in different leech species, owing to the shallow subdivision of each segment into three, five or more annuli. However, it is advisable to follow the traditional criteria of zoology, according to which 'true' segments are the units recognizable in the anatomy of the leech, in the nervous system especially. Back to numbers: the presence of exactly 32 segments could suggest that these are formed through a series of binary divisions, as $2^5 = 32$. This would be a simple and precise solution, but the body of a leech is not built this way. Along the posterior margin of the young embryo there is a transversal series of 10 cells that act as teloblasts – that is, a sort of fixed-term stem cells. Two cells with different destinies derive from the division of a teloblast: one of them keeps its teloblast character, while the other begins a path towards differentiation. In the case of the leeches' teloblasts, this process is repeated ca. 100 times. This generates 10 parallel strips of about 100 cells. Only one cell of each strip will become part of the pool of cells that will form a segment of the animal's body. Therefore, the process does not follow the simple and economic model suggested by the number of the segments eventually produced. First, the groups of cells that make up the 32 segments are not obtained by successive divisions of an initial pool. Second, the cells produced by teloblast proliferation are more numerous than those necessary and sufficient to give rise to 32 segments: the excess is therefore eliminated.

Body Axes, Appendages

Positions Along the Body Axes Are Specified During Morphogenesis

In February 1993, *Nature* published a short article that was soon acknowledged as a milestone in the history of understanding of developmental processes and their evolution. In one page, the authors, British biologists Jonathan M. Slack, Peter W. Holland and Christopher F. Graham, proposed the notion of the zootype, a sort of generalized model of animal organization, at least of animals with bilateral symmetry. In all of them, the main body axis is characterized by the early embryonic expression of a certain number of genes that are capable of identifying an antero-posterior sequence of positions. The names of some of these genes are a little esoteric, but in the past 30 years they have become very familiar to researchers involved in animal developmental genetics. According to our authors, the expression of two of these genes (*orthodenticle* and *empty spiracles*) identifies the front end of the embryo (and of the future animal); a series of domains follows, characterized by the expression of another seven genes, all belonging to the Hox family (p. 103): *labial, proboscipedia, Deformed, Sex combs reduced, Ultrabithorax, Abdominal-B* and *even-skipped*. A reference to *Drosophila* mutants is quite evident in the names of many of these genes: the *sex combs* are the 'combs' of sturdy bristles that characterize the front legs of male fruit flies; *proboscipedia* describes a mutant in which the 'proboscis' is replaced by a pair of legs; *Ultrabithorax* recalls the fruit flies that seem to have a double thorax, as they have two pairs of wings, whereas normal flies have only one pair.

The article contained a very important message: the expression of a small number of genes is sufficient to specify the antero-posterior polarity of the main body axis, along which, through the expression of these genes, a series of positions are marked. The different organs of the animal, therefore, occupy their own place by reference to an 'internal geometry' of the animal.

The article by Slack, Holland and Graham, however, did not apply only to *Drosophila*, or to insects, but to animals generally. This means that the mechanism by which positions along the main axis of the body are specified is universal, at least in principle. The result is a generalized model of animal organization, with a fore end that deserves to be called cephalic, whenever we find a brain and a mouth associated with it, and a trunk that continues to the rear end, where perhaps we expect to find the anus.

Research in the years that followed soon confirmed the general validity of the model. In the different groups, the genes involved in the patterning of the main body axis are more or less numerous, but these are always copies, differentiated to greater or lesser degree, of the genes listed a few lines above, or otherwise belonging to the same gene families.

However, these genes are responsible for patterning the main body axis, but not for its elongation.

In *Drosophila*, the two main body axes (antero-posterior and dorso-ventral) are already fixed during oogenesis. In *Caenorhabditis elegans*, as in many other animals, the primary spatial information for antero-posterior polarity is derived from a complex of microtubules that originates from an organizing core introduced into the egg by the spermatozoon. In some cases, the first indications of polarity are cancelled and subsequently replaced by new information. This happens for example in the case of polyembryony, as in *Copidosoma* (p. 43), and in embryos that undergo an 'anarchy' phase, such as the planarian referred to on p. 43.

Subsequent to the establishment of antero-posterior polarity in the egg, or – with the few exceptions mentioned – at a very early stage of embryonic development, the main body axis gradually lengthens, starting from the anterior end, and the different organs that occupy positions furthest from the front are built later than those that occupy front positions. But it is not always like this. In some insects, such as the fruit fly, both ends of the main body axis are already fixed in the egg, and the division of this axis into segments happens simultaneously rather than sequentially (p. 16).

A progressive elongation of the embryo, with the new elements added one by one at the rear end, is characteristic of many zoological groups, for example vertebrates. In modern scientific literature, this process is called 'terminal addition' – a term that requires no explanation but is questionable from a historical point of view. A century ago, terminal addition generally meant a developmental evolution model in accordance with Haeckel's theory (p. 71), according to which the development of an animal would summarize the evolutionary history of its species: a recapitulation lasting from a few weeks to a few years of a story that has unfolded over a billion years. An animal's ontogeny is compared with the (hypothetical) ontogeny of its ancestors, with respect to whom, according to Haeckel, a stage similar to that of the adult ancestor would be represented, in the descendant, by a juvenile stage; the latter would develop into adult through the terminal addition of a new final ontogenetic stage.

Returning to terminal addition understood in the current sense – that is, body lengthening localized at one end of the main body axis – the organisms that best approximate this behaviour are not animals, but fungi. In the following section we will deal with fungal growth, generally restricted to a pure terminal growth from the tip of a hypha. Different behaviours are shown by other filamentous organisms with a very simple morphology. For example, in filamentous green algae like *Spirogyra*, the main axis does not exhibit any obvious polarity, and elongation (in the prevailing vegetative stages at least) is diffuse, with any cell, apical or intermediate, having the same chance of dividing as any other cell in the chain.

Within a flower, the temporal succession according to which the different organs mature is very diverse: the outermost elements, the sepals, are not always the first ones to mature. Similarly, the first flowers that open in an inflorescence are not always those closest to the base; in willows, the first to open are often those most exposed to sunlight. The most curious sequence is found in the large flower heads of the wild teasel (*Dipsacus fullonum*), where the first flowers to open are neither the basal nor the apical ones, but a crown of flowers halfway; next to these, two diverging waves of floral opening move towards the two ends of the inflorescence. A similar bipolar progression is also observed in the development of the polyps of the colonial hydrozoan *Hydractinia echinata*.

Following Guidelines

Paths of Linear Growth Are Not Specified in the Genome

In addition to possessing chloroplasts, plant cells differ from animal cells because they are surrounded by a rigid cellulose wall. The walls of two contiguous cells are glued together, so the plant cells are denied any possibility of moving. Animal cells do not have walls, but this is not enough to allow them to move.

Ignoring the blood cells, which are suspended in the liquid plasma, is it really true that animal cells don't move? When a wound opens, the scar that soon closes it is produced by cells that move from nearby regions. Besides that, can a contribution of cell migration to morphogenetic processes be recognized, especially in the early stages of development? The answer is yes, for vertebrates at least.

In this animal group, an entire cell population that can be described as a fourth germ layer, in addition to ectoderm, mesoderm and endoderm, performs extensive migrations, at the end of which different subpopulations of cells are found in different parts of the embryo and will have very different fates. However, the topographical origin of these cells is common, and their collective name – neural crest cells – describes this. The neural crest is a complex of migratory cells that originates at the edges of the neural plate, which rise in the dorsal direction, converging until they are welded together to form the neural tube (the central axis of our nervous system). These cells, which initially look like motionless epithelial cells, undergo a transition to 'independent' (mesenchymal) cells and begin to move toward other, generally posterior, parts of the embryo. The length of their journey will vary (several groups, with different fates, can be distinguished), up to the different locations where they will stop and differentiate. The list of derivatives of the neural crest is very long. It includes, among others, elements of the skeleton (bone and especially cartilage), muscles, tendons, connective tissue, peripheral nervous system cells and even the cells that produce the iris pigment.

Extensive migrations, such as those of neural crest cells, are characteristic of vertebrates. In invertebrates, most cells form in their final position. But there are exceptions. In *Caenorhabditis elegans*, migrations of certain neuroblasts (neuron precursor cells) during larval stages span the entire length of the animal's body.

Often, it is not an entire cell that moves, but the growing apex of a very elongated cell that is otherwise immobile or anchored to neighbouring cells. Two important examples are provided by fungal hyphae and nerve cell axons.

When a fungus spore germinates, it generates a number of hyphae that spread radially, keeping each other at a distance. Apart from a singular case mentioned below, all vegetative hyphae (that is, those that are not part of reproductive structures) grow exclusively by extension at the tip. Many hyphae, however, are branched.

A good material on which to study fungal development is the red bread mould *Neurospora crassa*, which has long belonged to the list of model organisms. A colony of this mould consists of a network of hyphae in which an innermost region is recognized, where anastomoses between hyphae are frequent, and a peripheral area, with hyphae that branch but do not merge with each other: their reciprocal avoidance is a response to a chemical stimulus they emit from near the tip.

Growth polarity and functional polarity are not the same thing. Molecules inside a hypha can flow both in the direction in which the hypha grows and in the opposite direction. Nutrients enter the hypha at the tip and are transferred to the innermost parts. Other substances travel the other way round. We know many details of these processes thanks to studies on *Coprinopsis cinerea*, a kind of mushroom that develops on horse dung and similar substrates. In this environment, there are abundant populations of nematodes to which the mushroom would represent a good source of food. However, *C. cinerea* responds by releasing a variety of chemical weapons against the worms. A study of the transport of one of these molecules along the hyphae revealed surprising aspects. First, the signal travels at speeds of around 5 millimetres per second. Second, the fungus's response to the presence of nematodes travels only on a restricted class of specialized hyphae. Third, and this is the aspect that most interests us here, the flow is bidirectional: over time, phases in which the flow is directed towards the periphery alternate with phases in which the flow is towards the centre. A continuous reversal of polarity has been recorded in an organism very different from mushrooms – that is, in colonial ascidians, where blood flow alternates periodically.

Before abandoning fungi, let us take a look at the parasitic genera *Neotyphodium* and *Epichloë*, both of which develop as parasites inside the tissues of the host plants. In the diminutive spaces in which they live, the usual growth at the tip of the hypha is not feasible. These microscopic fungi grow in a different way, coordinated with the increase in size of the host cell: the hyphae grow in intercalary manner, that is, at various points along their length, and at the same speed as the host cell.

Strongly polarized cells that grow at one end, as mentioned above, also include the nerve cells that send out their axons to make contact with other nerve cells. Their path is highly disciplined, but it is not fixed genetically: in their lengthening, axons respond to molecular signals that come from other cells in the environment, which 'guide' them in their path. Some axons are several centimetres long, so the path they travel along can be more than a thousand times the diameter of the nerve cell of which the axon is a projection. This long road, however, is not completed in one breath, but in stages less than a millimetre long, which bring the axon into contact with a cell that acts as an intermediate finish line. This provides the axon with molecular information that directs it towards the next target. In addition, the wiring of the nervous system develops gradually. The first axons develop when the embryo is still relatively small and stretches out into a space where there are no other axons that can direct them, but things soon change. Most axons extend into an increasingly wider environment, where the axons that elongated before provide the younger axons with a track to follow until the next finish line, where they may move on to a different group of axons, which become their new guide. In humans, this process of axon-oriented stretching lasts several months.

Symmetry and Asymmetry

Symmetry Is Cheap, Deviations from Symmetry Are Expensive

In Sanskrit, its name is Dakshinavarti Shankh. It is the large shell of a marine mollusc living in the Indian Ocean, known in zoology as *Turbinella pyrum*. In the worlds of Hinduism and Buddhism, this is a sacred shell; it is often retouched, decorated with precious stones or metal inserts, and its natural apex is replaced with a mouthpiece that allows blowing during religious ceremonies. More precisely, the name Dakshinavarti Shankh is given to those rare specimens (less than one in a hundred thousand) in which the shell is coiled in the opposite direction to usual. To understand this better, let's put the shell in the standard position adopted in modern books on the subject: towards the top, we place the apex where the spiral coiling begins (the part of the shell produced by the young mollusc), while the opening of the shell faces us. In most snails, either marine, freshwater or terrestrial, the shell's opening is found on the right. These shells are conventionally described as right-handed. However, there are also left-handed species, in which the opening, with the shell in a standard position, is found on the left (Figure 7.2). All the species of a large family of land snails, the Clausiliidae, are left-handed: a widespread European species of this group is known as *Balaea perversa* because of its opposite coiling to the majority of snails. The fact that there are species with the opposite sense of spiral winding suggests that this character is not the result of chance and that it is probably controlled by genes. However, if we do not go beyond comparing different and perhaps unrelated species, it is difficult to address the problem properly. To do experiments, we need a species in which there are both right-handed and left-handed individuals, as in *Turbinella pyrum*.

One of these species is the freshwater snail long known as *Lymnaea peregra* (specialists today call it *Radix peregra* or *Peregriaria peregra*) which is much more suitable for experimentation than the sacred shell mollusc. Over much of Europe, it is easy to find specimens of this species in nature, and they can

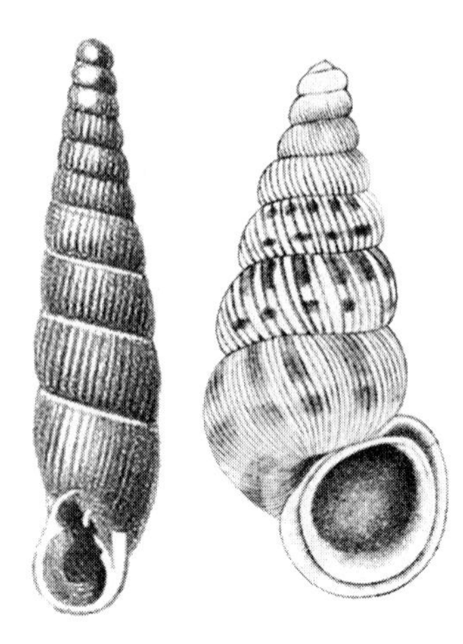

Figure 7.2 A left-handed shell (left, *Inchoatia haussknechti*, a representative of the Clausiliidae family) and a right-handed shell (right, *Cochlostoma septemspirale*).

easily be bred in a pot full of water with some aquatic plants which, in addition to keeping the water oxygenated, serve the mollusc as food. But there is more. The mollusc reproduces easily in captivity, and its egg-masses are almost transparent, thus making it quite easy to observe embryonic development. The first indication of what will be the direction of coiling of the future shell is already evident during the third cell division (Figure 7.3).

A hundred years ago, the Swiss scholar Jean Piaget, better known for his later contributions to psychology, showed in *Lymnaea stagnalis*, a close relative of *L. peregra*, that the orientation of the blastomeres (and the future coiling of the shell) is genetically controlled: however, it does not express the genotype of the individual itself, but the genotype of the mother. This seemed to indicate a transmission of information through the cytoplasm. This hypothesis has been confirmed. The maternal genotype determines the orientation of submicroscopic fibres, the so-called cytoskeleton, in the eggs that the animal produces. The orientation of the cytoskeleton, in turn, influences the orientation of the blastomeres during the early stages of embryonic development, giving the new individual a 'switch' between left and right that it will carry throughout its life and that will become clearly visible in its shell.

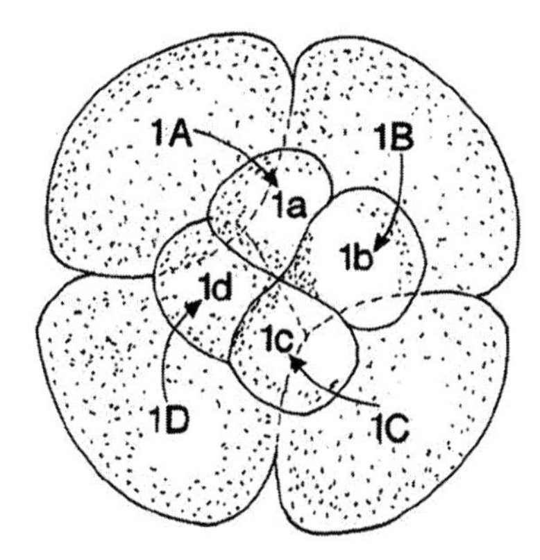

Figure 7.3 Schematic drawing of the eight-cell stage of an embryo with spiral cleavage, as typical of molluscs, annelids and other invertebrates. At this stage, the embryo includes four larger cells (macromeres) and four smaller cells (micromeres). Blastomeres labelled with the same number, for example 1A and 1a, are sister cells – that is, they derive from the division of the same cell (in this case, A). Besides the contrast between macromeres and micromeres, spiral cleavage is characterized by the reciprocal position of the two quartets suggestive of a 'spiral' twisting.

An experimental verification was obtained a few years ago. By inserting glass rods between the blastomeres, a researcher succeeded pushing the cells in the opposite direction than expected according to maternal inheritance. This manipulation turned dextral (right-handed) embryos into sinistral (left-handed) ones and vice versa. The molluscs that survived this treatment later developed shells coiled according to the forced direction rather than to their original one. Research in recent years, conducted on other kinds of molluscs (the right-handed marine limpet *Lottia gigantea* and the left-handed freshwater snail *Biomphalaria glabrata*), has led to the identification of a gene that codes for a molecule (formin) that acts as a regulator of the architecture of the cytoskeleton and therefore, indirectly, of the direction in which the shell is coiled.

Transforming a right-handed shell into a left-handed shell, or vice versa, is not possible. Once the development of the animal has moved away from the symmetrical condition in a given direction, the only way to transform one of the two variants into the other is to use a mirror. However, a right version and a left version of the same part of the body coexist in most animal species, including humans. A plane of symmetry, in fact, divides our body into two halves which, in their external aspect and apart from minor details, are the mirror image of each other. Of course, not even in a strictly deterministic vision that sees development as the execution of a programme written in the genes should we think that there are separate 'instructions' for the realization of the left hand and the right hand, the left ear and the right ear.

Genes, however, are involved in deviations from symmetry, both in the case in which they affect the whole body, as in snails (including the shell), and in the case in which the asymmetry concerns single organs or complexes of organs. Here we can return to the human body (although a chick or a frog would be equally good), where external symmetry is matched by a striking asymmetry of the viscera. Just think of the position of the liver, heart or stomach in the human body. Another similarity to what we have seen in snails is that even in humans there are individuals in which the orientation of the visceral mass is opposite to that of the great majority of the population: individuals, that is, in which the heart is displaced to the right, the liver to the left. This anomaly can affect the whole of the viscera or individual organs only, while the others are in the usual position. In humans, the causes of this symmetry reversal (technically, *situs viscerum inversus*) are not always the same. In many cases, it is due to an initial asymmetry manifested in the embryo at the so-called primitive node, about the third week after fertilization.

So far, we have mentioned only animals. But left–right asymmetry occurs in plants too. We can easily see this in the shape of the leaves of a great many plants, such as in lime trees (*Tilia* spp.) or hazel (*Corylus* spp.), and in the winding direction of climbing plants. In some instances, handedness is a stochastic effect, in others it is under genetic control. The latter is likely in those twining plants that are consistently either right-handed, such as honeysuckle (*Lonicera* spp.) and hop (*Humulus lupulus*), or constantly left-handed, such as bindweeds (*Convolvulus* spp.).

Temporal Serial Patterns

Sequentially Produced Equivalent Parts Are Not Necessarily Mere Repetitions of an Elementary Module

In a previous chapter (p. 85), I devoted a few pages to regeneration, but I did not address in detail the circumstances that may cause an animal to lose a body part. We might take it for granted that loss is caused by accident, including the unfortunate circumstance of growing up in a developmental biology lab. But there are also animals that lose a body part through natural causes. These can be in response to the attack of a predator, with whom it may be better to leave only a tail or a limb (while fleeing in the meantime); or even a physiological event, perhaps seasonal, such as the loss of antlers by a male deer. In these cases, the loss of an organ is but a special aspect in the development of the individual, which takes the name of autotomy.

Autotomy is often followed by regeneration, although in many cases the regenerated part looks somewhat different and, frequently, smaller than the lost part. Sometimes, however, autotomy cannot be followed by regeneration: this is the case for some insects (the bugs of the coreid family) that can autotomize a limb when adults – but moults and regeneration are not possible in the adult insect.

Echinoderms, in contrast, are champions at autotomy followed by regeneration. In sea cucumbers attacked by crabs, the digestive tract and other viscera are literally spilled out through the mouth and, after being lost, are subsequently regenerated.

Reluctance to regard the loss of a body part as a topic in developmental biology is understandable. But perhaps we should be aware that autotomy takes place along preformed rupture lines and in ways that allow damage, such as the loss of blood and other internal liquids, to be kept as low as possible. Sometimes the loss of a body part is a regular phenomenon that affects all individuals of the species and occurs at specific times in their life. Examples are the fall of male deer antlers and the physiological loss of teeth in vertebrates.

In the case of deer, the new antlers produced at the beginning of the new year are not a precise copy of those lost the previous year, but are larger and often equipped with a higher number of prongs. Every year, the entire structure is built from scratch, so the old one is not available to serve as a scaffold for the new one. The new antlers are thus the product of a whole morphogenetic process repeated year after year with minor variations.

In mammals, deer antlers are the only organ that an animal can completely renew, and in a very efficient way. The antlers of the European red deer grow at a rate of up to 17 millimetres per day, supported by cell proliferation at least as fast as in cancerous tissues. Interestingly, the incidence of cancer in deer is much below the average for mammals.

Thus, the antlers a deer proudly displays this year are not the same as last year's, but a new version of the same body part, built at a different date. Therefore, as we describe the vertebrae of the same animal as structures in serial homology, so we can describe the antlers produced by the same individual in subsequent years as organs in *temporal* serial homology.

In the case of vertebrate teeth, the two aspects generally coexist: serial homology between the teeth an individual possesses at a certain time, and temporal serial homology between a tooth and the one that replaces it when the first is lost.

Teeth are a good example of modulation upon a theme. In some vertebrates (including in some mammals, such as dolphins), all teeth are very similar to one another; in others, however, there are obvious differences along the series. This is evident in humans, where incisors, canines, premolars and molars are distinguished, but much more evident in those animals, such as narwhal and elephants, in which one or two teeth (the tusks) grow to a much greater size than the others. Even along the time series there may be differences, not only in size: that is, the next tooth can have a different shape from the one it replaces.

The extent and distribution over time of the events leading to tooth renewal are very different among different vertebrate species.

In sharks, which are almost always equipped with numerous rows of functioning teeth, some of these are continuously lost, while new rows of teeth progressively grow to full size. The small cigar shark (*Isistius brasiliensis*), however, has a single row of teeth (about 30 teeth on the upper jaw, almost as many on the lower) and renews them all together.

In many vertebrates, the teeth that are renewed at a given time are not contiguous with one another, but alternate, a mechanism that guarantees the animals always to have enough teeth in a functional state.

In some species of fishes, including swordfish, and in the platypus, teeth are present in the young and eventually lost but not replaced, so mature individuals have no teeth. In most mammals, the first teeth (the 'milk teeth') are replaced, but only once; some, such as kangaroos and elephants, replace their teeth several times. The same happens in different species of lizards and snakes: venom teeth are replaced up to eight times. Alligators, the record holders, replace teeth up to 50 times.

Fractals and Paramorphism

Each Tier in the Hierarchy of Developmental Modules Is Not Specified Independently from the Others

In close-up, the lumpy surface of the cauliflower resolves into a multiplicity of bulges sprinkled with 'warts' arranged in multiple spirals. Focusing on a detail, we see the same pattern as in the entire cauliflower, repeated on a different scale.

This presence of a structural model, identically repeated at different scales, is characteristic of a class of mathematical objects – fractals – that became a fashionable subject around 1980, when the Polish-French mathematician Benoît Mandelbrot developed this chapter of geometry. Examples taken from the plant world (in addition to cauliflower, for example, the fronds of many fern species) or from animals (for example, the suture lines of many ammonite shells) are still among the most popular of the real objects with a fractal structure (Figure 7.4). Of course, in a fractal as a geometric entity the basic pattern is repeated *ad infinitum*, while natural objects with a fractal structure repeat the structural model only a small number of times – the coarseness of the material they are made of does not allow for more.

The resemblance of some biological objects to fractals is very attractive from the perspective of a developmental biologist, because it suggests the possibility (to be demonstrated, however, by facts) of producing complex structures through a process that includes only the production and the iteration of a simple structural model. In the language of computers: just give instructions to generate the elementary pattern, plus some instructions for triggering its iterations. It would be simple and elegant if plants and animals were built like this. However, as we saw on p. 113, it is not always true that living beings develop according to the sequence of stages that would be the most direct and simplest in our abstract models, inspired by

Figure 7.4 The outline of the leaves of *Macleaya* (here reproduced from a photograph) matches very closely the pattern of one of the most popular fractal objects, von Koch's (or snowflake) curve.

mathematical elegance. For example, fractal geometry may seem fitting in a description of the branching pattern of a fruit fly's tracheal system, but the first-, second- and third-order branching are actually under different genetic control.

Nevertheless, even a less precise correspondence between the parts of an organism may suggest the existence of developmental processes that (perhaps with minor adjustments) can produce parts of the body that at first sight do not seem to be closely related in morphology. To 'see' these correspondences, a philosophical mind can be helpful.

In 1851, an article appeared in the *Transactions of the Botanical Society in Edinburgh* with the following rather uninformative title: "Some remarks on the plant morphologically considered." The author was Reverend James M'Cosh, a Scottish philosopher who had recently become professor of logic and metaphysics at Queen's College (now Queen's University), Belfast; he moved later to the United States, where for 20 years he was president of the College of New Jersey, now Princeton University. When the *Origin of Species* was published, he attempted to reconcile religion and evolutionary theory. In his 1851 article, M'Cosh stated the profound unity of the structural design of the plant. In support of his thesis he presented a long list of facts, from which I take some examples:

> According to our idea, [a plant] consists essentially of a stem sending out other stems similar to itself at certain angles, and in such a regular manner, that the whole is made to take a predetermined form. The ascending axis for instance sends out at particular normal angles in each tree, branches similar in structure to itself. These lateral branches again send out branchlets of a like nature with themselves, and at much the same angles. The whole tree with its branches thus comes to be of the same general form as every individual branch, and every branch with its branchlets comes to be a type of the whole plant in its skeleton and outline.
>
> Taking this idea of a plant along with us, let us now inquire whether there may not be a morphological analogy between the stems and the ribs or veins of the leaf. As these veins are vascular bundles, proceeding from the fibro-vascular bundles of the stem, they may be found to obey the same laws.
>
> Some trees [. . .] send out side branches along the axis from the root, or near the very root, and the leaves of those trees have little or no petiole or leaf stalk, but begin to expand from nearly the very place where the leaf springs from the stem. There are other trees [. . .] which have a considerably long unbranched trunk, and the leaves of these trees will be found to have a pretty long leaf stalk. [. . .]
>
> Generally we shall find a correspondence between the angle of the ramification of the tree, and the angle of venation of the leaf [. . .]. All that we argue for is a general correspondence between the tendency of the direction of the branches, and the tendency of the direction of the veins of the leafage.

These pages have remained practically ignored to the present day, but a structural relationship between stems and leaves was formulated again around the middle of the last century by Agnes Arber, a British plant morphologist, historian of botany and philosopher of biology.

In fact, this kind of structural correspondence is even more pervasive than emerges from M'Cosh's pages. For example, in plants with pinnate leaves there is often agreement between the arrangement of the leaflets within the

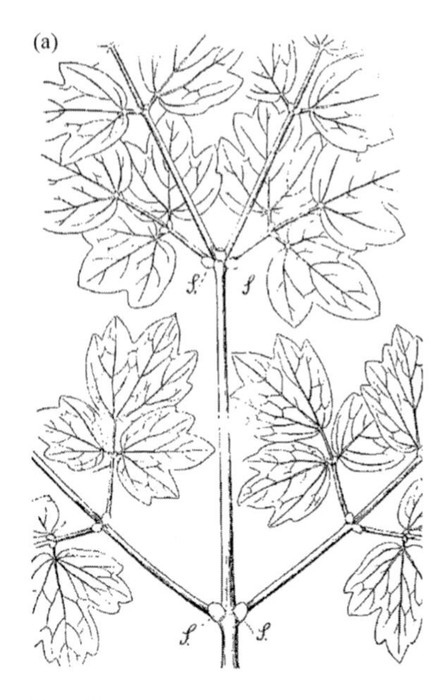

Figure 7.5 (a) Paramorphism in plants: in many plants with compound leaves, as in this *Thalictrum aquilegifolium*, the pairs of leaflets forming the leaf are inserted opposite to each other on the leaf's rhachis (its main axis), and the whole compound leaves are also inserted opposite to each other on the plant's stem. (b) In other plants, as in this *Anthyllis vulneraria*, leaflets are inserted individually on the leaf rhachis, as are the leaves on the plant's stem.

compound leaf and the type of phyllotactic distribution of the compound leaves along the stem. There are, for example, plants with opposite (technically, 'decussate') compound leaves, such as elder (*Sambucus*) and ash (*Fraxinus*) species, and plants with compound leaves attached singly to the stem, such as the black locust (*Robinia pseudacacia*). In the former, the leaflets are arranged within each compound leaf in a very orderly manner, in pairs of opposite elements; in the latter, the alignment is inaccurate, and each leaflet seems to occupy an independent place within the compound leaf (Figure 7.5).

Are there similar regularities in the animal kingdom? Yes. Many examples are found in animals with appendages. The latter are sometimes divided into

(b)

Figure 7.5 (*cont.*)

segments, like the legs of insects, scorpions (Figure 7.6) and centipedes, but others are not, for example the tentacles of octopus and snails. It is easy to note that the main body axis is also divided into segments in insects and centipedes, but not in octopus and snails. This is just one of the many examples of the structural correspondence that often (but not always) occurs between the main axis of an animal's body and the axis of each of its appendages. This correspondence has been called paramorphism.

In the different groups of centipedes, the correspondence between the main axis and the appendages extends to the way in which these axes develop during the post-embryonic life. In some centipedes, including scolopenders, at the time of hatching all body segments are already fully developed and the

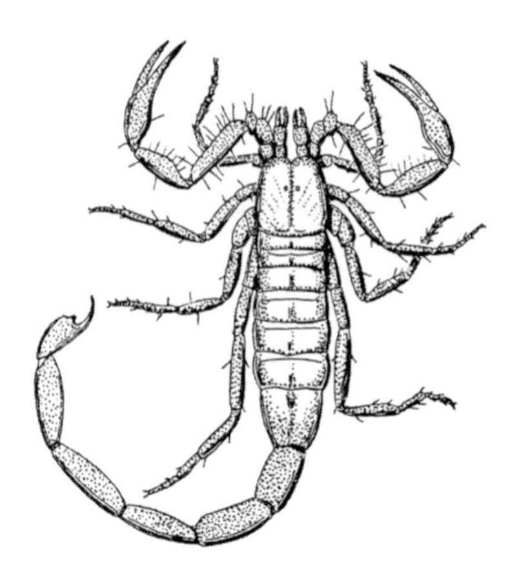

Figure 7.6 Paramorphism in animals: segmented appendages in a segmented animal (scorpion).

antennae are already articulated with their final number of joints; in some other centipedes, the individual hatches with an incomplete number of segments and the same applies to its antennae.

If appendages, segmented or not, are a very common feature of animal morphology, copies of the main body axis grafted on the sides of the original axis are probably unexpected, except in the domain of teratology. However, a marine annelid worm with lateral branches (*Syllis ramosa*) was described as long ago as 1879, and a second species (*Ramisyllis multicaudata*), belonging to the same family (Syllidae), has been described recently (2012). The latter worm, with its extensive and seemingly indeterminate system of hundreds of branches of many orders, is by far the most impressive example of paramorphism in the whole animal kingdom.

8 The Ecology of Development

The Leaves of a Bonsai: Developmental Robustness

Developmental Plasticity Is Not Necessarily Alternative to Developmental Robustness

In the insect collection I formed in my youth, a small male specimen of the longhorn beetle *Aromia moschata* (a species with a splendid green-blue metallic livery, whose larvae develop in the woody tissue of large willows) stood out: its length was only two-thirds that of the other specimens of the species, so its weight must have been no more than a third of normal. This beetle had nevertheless managed to develop to the adult stage.

Entomologists are well aware that the adult size of some beetles can vary greatly even within a local population. These insects develop as larvae in confined spaces, inside a substrate that provides them with nourishment and protection, but is also equivalent to a prison from which they cannot escape before having completed metamorphosis, if they succeed in growing up to that stage.

If we evaluate this story from the perspective of survival, we must recognize that the ability of these insects to make the most of an environment that offers no alternatives is truly remarkable. However, this is a reading in terms of evolutionary biology; what is the implication of the same story for developmental biology? Compared with normal-sized specimens of its species, the small *Aromia moschata* in my collection did not present perceptible morphological differences: at most, a slightly different relationship between the length of the antennae and the length of the body, or a slightly dissimilar surface sculpture of the elytra. It is probable that even an examination of the internal anatomy would not have revealed significant differences.

This case is instructive, but how representative is it of the developmental responses of an organism to exposure to different environmental conditions? For simplicity, let's compare conspecific individuals with no genetic differences between them, at least no differences capable of modifying their developmental response to different environmental conditions. In this context, the beetles that develop in the woody tissues of trees are an example of developmental robustness, an aspect we will explore in this section. Other organisms, if exposed to different environmental conditions, develop into alternative phenotypes: that is, they manifest phenotypic plasticity, discussed in the next section.

Developmental robustness is defined as the ability of an organism to produce its targeted phenotype regardless of perturbations caused by internal or external factors; but a form of developmental robustness operates even when the external and internal environmental conditions are favourable. In many plants, the sizes of the individual flower parts and also those of the leaves are fairly uniform, even if the number of cells is far from constant. In other words, we are not in conditions of eutely (p. 23). There are no mechanisms, here, for counting cells or evaluating their size. Apparently, robustness depends in this case on compensation mechanisms, whereby larger cells are produced in organs where mitotic activity has been slower or ended earlier: the result is the production of organs of fairly uniform size.

We must admit a certain degree of arbitrariness in any assessment of the robustness of an organism. Our usual tendency to evaluate phenotypes from the point of view of their adaptive value can lead us to consider the developmental trajectory of an organism robust if it does not significantly reduce its ability to reproduce.

At the edge of roads, in a narrow strip of land where environmental conditions are often difficult for the plant development (scarcity of humus, frequent drying, accumulation of toxic substances washed away from the road), small plants are often observed with phenotypes known in the German botanical tradition as *Hungerformen* – starvation forms. It is easier to notice their presence when these are dwarfed and stunted individuals of species that usually produce showy flowers, such as poppies. Even starvation forms, indeed, often manage to produce flowers, perhaps even to bear fruit. This can be seen, for example, in some bonsai trees. In addition to their overall size, they often differ from normal individuals in other characters, but not always the same ones. In some species, changes are more evident in the size of the leaves, flowers or individual parts of the flowers, sometimes in their number, sometimes in what we can call the plant's temporal phenotype, its flowering calendar especially. The most extreme phenotype ever described for a plant *Hungerform* is the diminutive specimen of the genus *Littorella* (a relative of fleaworts, *Plantago* spp.), described by Agnes Arber, that managed to flower in spite of the adverse environmental conditions (drought): it demonstrated its robustness in the production of stamens of normal size, whereas the rest of the plant was shorter than the stamen filaments.

There are limits to robustness, of course. Sometimes, an animal's development is dramatically diverted by a parasite. The victim's reproductive system often suffers most. I have already mentioned (p. 48) castration induced by *Sacculina* on parasitized crabs. In a male crab, this translates into becoming superficially similar to a female. In other cases, the effect of the parasite is very selective, limited to individual organs. For example, in *Chironomus* midges, males parasitized by nematodes have normal genital appendages, typical of their sex, but their secondary sexual characteristics suffer: the forelegs are not as long as they should be in males, and the antennae are simpler than normal and resemble those of females.

Developing in Alternative Environments: Phenotypic Plasticity

Alternative Phenotypes, Without Intermediates, Do Not Necessarily Depend on Genetic Differences

Among the organisms that reproduce only sexually, there are no two individuals with identical genotype, except for monozygotic twins. However, many differences between one individual and another, within a population, can have causes other than genetic differences. In particular, the conditions in which two individuals have developed can have conspicuous consequences on the size and morphology of adults. In the previous section I remarked that differences in size between individuals may depend on conditions encountered during development. Much more interesting than size, however, are the qualitative differences between distinct classes of individuals such as males and females or, to remain in the context of insects, winged and non-winged adult individuals, regardless of sex. Or workers and queens, if they are bees or ants; sedentary and solitary individuals or gregarious and migratory, if we are dealing with certain species of locusts. In these cases, the fact that an individual develops as a male or female, winged or wingless, and so on, suggests a rigorous control mechanism, such as we can only expect from genes. Is this always the case? As we will see in the following lines, the answer is no. But we will also see how thin can be the divide between differences that are controlled by genes and differences that are not.

Let us start by introducing two terms: polymorphism and polyphenism. *Polymorphism* is the presence, in a population, of two or more distinct classes of individuals, without intermediates, dependent on precise differences in one or more genes; *polyphenism*, instead, is the presence, in a population, of two or more kinds of individuals, the difference between which depends on the environmental conditions in which development took place, at least in a critical, sometimes short, phase. We might expect that in polymorphism the differences between individuals of the two forms are more conspicuous or more constant than in polyphenism, but this is not necessarily the case. Even the two sexes, male and female, do not always have a genetic or chromosomal basis. In several reptiles, for example, the sex of the individual depends on the temperature to which the embryo was exposed during a sensitive phase of incubation. How little this mechanism is deterministic can be seen from the fact that in some species (such as the American alligator), development goes in the male direction when the embryo is exposed to lower temperatures than those that determine development to female, whereas in other species (such as the loggerhead sea turtle *Caretta caretta*), lower temperatures lead to female, higher temperatures to male. In the case of locusts, the colour and behaviour of the adult correlate with the degree of crowding experienced by the insect at the beginning of its active life: young individuals that develop in conditions of low population density become solitary and sedentary adults; those that grow in crowded conditions and are very often in physical contact with other individuals become gregarious and migratory adults.

In plants, different phenotypes can coexist within the same ramet (p. 40), a proof of how problematic it is to describe plants as individuals. The presence of alternative phenotypes in the same ramet arises because the production of a new part, under new environmental conditions, has no consequences on the definitively fixed morphology of the parts that developed previously.

Under different environmental conditions, several plants produce two very different types of leaves, a phenomenon known as heterophylly. One of these species is the lesser spearwort *Ranunculus flammea*: the leaves of this aquatic buttercup that develop above the water surface are broad and lanceolate, while the leaves that develop in the water are linear.

Among the many species with conspicuous polyphenism, one of the most interesting (and best studied) is the pea aphid. In this small insect, both males and females come in two distinct forms: with or without wings. The nature of the phenomenon, however, is different between sexes. In females, it is polyphenism: wings develop, or not, depending on the environmental conditions to which the mother was exposed when embryos were developing in her ovary. In males, instead, the difference between winged and wingless individuals is genetic in nature: it is a polymorphism based on a gene (*aphicarus*) located on one of the sex chromosomes. The divide between the polyphenism of females and the polymorphism of males is very subtle, as the product of the gene locus controlling wing development in males is also involved in the response of females to environmental cues.

The effects of phenotypic plasticity are often irreversible, but sometimes are reversible.

An example of irreversible polyphenism is provided by the moth *Nemoria arizonaria*, the caterpillars of which grow on oaks. This species completes two generations per year. Those of the spring generation are active in the season when the tree bears catkins (male inflorescences); these are no longer present when the caterpillars of the second generation develop. The larvae of this moth have mimetic colours, conspicuously different between generations: those of the spring generation resemble catkins, those of the second resemble the twigs of the oak. Greene's experiments have shown that this difference is the result of a difference in diet – poor in tannin for the caterpillars of the first generation, which feed on catkins, and rich in tannin for those of the second generation, which feed on leaves.

An example of reversible polyphenism is the sexual phenotypes of fishes such as the small goby *Trimma okinawae*. This species lives in groups within which there is a hierarchy led by a dominant male. If this is removed, its place is taken by the highest-ranking female, who changes sex. Conversely, in the presence of a larger male, a smaller male can change sex in the opposite direction, becoming a female.

Developmental Timing

Phenotypes Are Not Necessarily About Morphology

In the previous pages, we have seen many processes through which the phenotype of an animal or a plant changes throughout development. However, an important aspect was missing. The changes we have dealt with are almost exclusively about shape (and, somewhat marginally, size), but the phenotype of a plant or animal is not restricted to that. There are additional aspects, neglecting which amounts to ignoring entire chapters of developmental biology. Or potential chapters, perhaps, because these subjects are not usually discussed in books and journals of developmental biology, but not for any good reason.

The aspect to which I wish to draw the reader's attention is the temporal organization of the animal or plant. For any mammal, for example, there are different aspects of biology that have a characteristic duration, or occur approximately at a certain age. Among these are the duration of gestation, the age at which weaning occurs and the age at which sexual maturity is reached. The fact that these times or durations are subject to individual variability and that, whatever the control that genes exert on them, they are subject to environmental influences, is not a sufficient reason to ignore them, or to contrast them with the morphological aspects of the phenotype: temporal aspects too are dependent on both genes and environment.

In the following paragraphs, we will see examples of particularly dramatic temporal phenotypes. In most cases, we know very little about the mechanisms that determine their accuracy. This is not a good reason to neglect them; on the contrary, I think it is a good stimulus for research to take them into consideration in the context of developmental biology.

The most sensational example from the animal kingdom is probably the length of the life cycle of some North American cicadas (genus *Magicicada*). Like all cicadas, they spend most of their lives underground, gradually increasing in size through five nymphal stages. When a nymph is ready for the last moult, it emerges from the ground. The adult spends its short existence in the trees. The sounds produced by *Magicicada* are loud to the point of being dangerous to the human ear, especially because these cicadas complete their biological cycle with surprising synchrony. Depending on the species, the whole development takes 13 or 17 years. A year in which the woods resound with their deafening chirps is followed by 12 or 16 years of silence, interrupted only by a small number of individuals whose timers are out of sync.

Less conspicuous, but much more generalized, as it involves millions of different species, is the succession of moulting in the different groups of arthropods. This number is small in most insects (often four or five) but can be up to a few dozen. As expected, when the number is higher, in general it is also more variable, sometimes even between individuals issuing from the same parents. But very often it is fixed, even if the conditions in which the individual grows up affect the duration of individual stages.

The plant kingdom is literally full of temporal phenotypes. Overall, we know much more about their production than the little we know about most temporal phenotypes in animals. Particularly studied are the mechanisms that determine the flowering season, often (but not always) in relation to the photoperiod (the relative duration of the hours of light and dark in a 24-hour cycle).

The distinction between annual, biennial and perennial plants is also one of temporal phenotypes. Annual plants develop, grow, bloom and bear fruit within a total span of a few months, and then they die. Biennial plants have a first year of vegetative growth, while postponing flowering and fruiting to the second year. Perennials live longer, up to a few hundred years – even more than 4000 years in exceptional cases like the Great Basin bristlecone pine (*Pinus longaeva*) and the giant redwood (*Sequoiadendron giganteum*) – but they begin to bloom only after a juvenile phase that can last several years. Their reproductive activity is often repeated every year, perhaps with alternating years of greater abundance and years in which flowering is more modest. Some perennials, however, bloom only once in their life (p. 149).

Like many other phenotypes, morphological and temporal, the distinction between annual, biennial and perennial plants is not strict. In some species, such as the marsh yellow cress (*Rorippa palustris*), there are annual, biennial and perennial individuals – an ideal situation in which to study the mechanisms that regulate these temporal phenotypes, but we must be careful not to suppose that what is valid in this species applies to all flowering plants.

The same caution is necessary when, based on studies carried out on a few species, we want to explain how the time over which one individual flower remains open is controlled. In the orchid family alone, this duration varies, depending on the species, between 5 minutes and 9 months. Another question: how long does it take for a leaf, already recognizable on the plant, to reach its final size? This time can be very long in cold and difficult environments such as Arctic or alpine zones. A leaf of Alpine bistort *(Bistorta vivipara*, also known as *Polygonum viviparum*), a species of those environments, takes 5 years to unfold. But environmental conditions do not explain why it also takes 5 years to unfold a frond of the cinnamon fern (*Osmundastrum cinnamomeum*), which inhabits marshes in temperate and tropical regions.

All these temporal phenotypes are expressions of developmental processes in which, as is also the case for morphology, the roles of genes and environment are intertwined. In the small model plant *Arabidopsis thaliana*, over 300 genes have been identified with effects on the flowering time.

Periodization of Development

Developmental Phases Are Not Necessarily Well-Defined Steps Along the Individual's History

Signs of incoming puberty in humans are unequivocal. The first beard hairs begin to appear on the face of a boy, and his voice becomes lower and deeper. The long transition from childhood to adulthood is taking place. But this step did not begin on a precise day, marked by sudden and clearly recognizable changes, nor will it be complete on a precise day. Although puberty is a stage of development in which specific changes take place in the individual, its limits are uncertain, and its duration is only predictable in a very rough way.

This is an example of the difficulty we often encounter in dividing development into phases – that is, in giving it a periodization. There are very few fixed points that seem to offer a precise reference: the date of birth, perhaps, and, for females, the menarche.

Strict periodization, however, would be very useful as a support for experimental work. How can we evaluate, for example, the results of an experimental treatment, if we cannot safely identify the treated animals and those we consider as a control group as belonging to the same developmental stage? This problem has long been addressed by embryologists, who have produced standard development tables for a number of model animals. Such tables are available, for example, for *Drosophila melanogaster*, *Caenorhabditis elegans*, zebrafish and mice. Theiler's *Atlas of Mouse Development* recognizes 25 prenatal stages, while Campos-Ortega and Hartenstein divide the embryonic development of the fruit fly into 17 stages.

Periodizing post-embryonic development seems to be a more difficult undertaking, but not for all animals. In a whole phylum – indeed, in the most species-rich of all animal phyla, the arthropods – post-embryonic development is marked by a series of events that can help in articulating development into temporal units. Arthropods undergo moults, often the same number or nearly so in all individuals of a given species. Between one moult and the next, the animal cannot change shape, because it is covered by an inextensible external skeleton. The only exceptions to this otherwise strict rule are those granted by the flexible membranes that connect the individual pieces of the exoskeleton. These membranes are very extensible in some female insects, such as the queens of some termite species, a condition very favourable to raise their fecundity to astonishing levels.

A moult can separate two instars that are similar apart from size, which as a rule increases from one stage to the next; but it can instead be associated with important structural change, in which case we say that the animal is undergoing metamorphosis.

In insects like *Drosophila*, the periodization of the post-embryonic development seems to be uncontroversial. At egg hatching, the insect is in the stage of larva I; two moults will make it progress to larva II and larva III, respectively. Two more moults will follow: the first of these leads to the stage of pupa, the second to the adult. There will be no further moulting: in *Drosophila*, as in almost all insects and in many other arthropods, the adult does not moult.

This periodization is a convenient system of reference when we need to specify the stage of post-embryonic development on which we are doing observations or experiments: moreover, it allows easy comparisons between species. But there are problems.

To begin with, it is all too easy to trust numbers more than these deserve. For example, if a plant bug and a mosquito perform the same number of moults (five) throughout their post-embryonic development, is it legitimate to consider the second post-embryonic stages of the two insects to be equivalent? There are good reasons (widely discussed in the specialist literature) that force us to answer negatively. It is enough here to mention that in the second post-embryonic stage the plant bug is a young nymph largely similar to a small, wingless version of the adult, whereas the second post-embryonic stage of the mosquito is an aquatic legless larva fated to undergo a radical metamorphosis. Still on the subject of numbers, there may also be more subtle problems, perhaps of little practical interest, but not negligible, because they challenge the solidity of our periodization of development.

In many butterfly species, for example, the number of moults is not the same for all individuals. For example, in the large white (*Pieris brassicae*), the number of larval stages varies from four to six, depending on the temperature. If two individuals of this butterfly species develop through different numbers of larval moults, how do we compare them? For example, is the fourth larval stage of the first individual equivalent to the fourth larval stage of the other, even if for one specimen the fourth is the last larval stage, to be followed by the pupa, while in the other case another larval stage will precede the pupa? The curious reader may explore the technical literature (p. 172). Here it is convenient to shift the attention to a more serious and more general difficulty.

When we periodize the post-embryonic development of an arthropod based on moults, we accept two important points without discussion: first, that the starting point of post-embryonic development is the same for all arthropods; second, that no significant event occurs between one moult and the next. Both assumptions are far from granted.

The events surrounding egg hatching are a grey area in the development of arthropods. By the time it leaves the eggshell, a young animal is often complete with all its body segments and appendages, and is able to start active life immediately, but in many arthropod groups it still looks embryonic. In many spiders, for example, when the eggshell (chorion) is broken, the animal is a motionless pre-larva, with legs still incompletely articulated. It will take it a moult or two to develop into the first mobile instar (nymph I).

In many insects, such as butterflies, flies and beetles, the first instar is immediately active, but in others such as dragonflies and grasshoppers, the first instar that we formally describe as post-embryonic, because it becomes visible when the chorion splits, is a pre-larva that will undergo a moult before starting active life.

What happens among centipedes is particularly intriguing. When the egg hatches, some centipedes – brown centipedes (lithobiomorphs) and house centipedes (scutigeromorphs) – are already able to go around looking for food; but others – scolopenders and geophilomorphs – look embryonic and will start active life only after completing a couple of moults. However, from a different perspective, scolopenders and geophilomorphs seem to be at a more advanced stage of development, since by the time of hatching they already possess all the segments (often very numerous) characteristic of their species. In contrast, brown and house centipedes, although immediately active after hatching, begin post-embryonic development with numbers of segments and pairs of legs much lower than the definitive one, to be reached only after a few moults.

In summary, at the time of hatching, arthropods are not all at the same degree of morphological differentiation. Hatching is thus a questionable point of reference for comparisons of the developmental stages of arthropods.

The divide between embryonic and post-embryonic development is also uncertain in animals other than arthropods. A critical case – and a decidedly unusual one – is once more that of planarians. During embryonic development, we expect these animals to feed on the yolk provided by the mother, but we certainly do not expect the embryo to use a pharynx. However, this is the case in a common European planarian (*Dendrocoelum lacteum*). After an initial phase of blastomere proliferation, an embryonic pharynx and an embryonic intestine differentiate very early. Before the remaining organs take shape, the embryonic pharynx begins to suck the yolk stored in a syncytial tissue surrounding the embryo.

If the periodization of development in flatworms does not seem interesting enough, let's move to a group of animals much closer to us, the marsupials. Where are we going to fix the dividing line between the embryonic and post-embryonic phases in the life of a kangaroo: at the time the offspring, currently weighing just 30 millionths of the weight of the mother, is transferred from the uterus to the pouch, or only at a later stage?

Stipulating Age

Age from Fertilization Cannot Always Be Objectively Determined

Developmental biology can contribute a lot of evidence to the debate on a fundamental bioethical question: when, precisely, does a human individual begin its existence? Answering this question, however, is not up to biology, which can only flag important events in the rich periodization of human embryonic development. Among these events are the implantation of the blastocyst in the uterus, the beginning of the transcription of the zygotic genome, and the achievement of some level of complexity in the organization of the brain and in the manifestation of its functionality. Furthermore, biology can flag – in humans or in other animal species – particular developmental events that make uncertain not only the determination of a threshold beyond which the presence of a new individual is irreversibly fixed, but even the age of the individual.

Of all the events listed in the previous lines, the most visible is of course birth: for this event, it is objectively possible (and generally not problematic) to accurately fix the date that ends up on the person's identity document. The date of birth, rather than the presumed date of conception, fixes a person's identity in astrology too. In this regard, however, a different perspective may be inspired by a page by Ptolemy, the Greek mathematician, astronomer (and astrologer) of the second century of our era, who dedicated his work *Tetrabiblos* (The Four Books) to the alleged influence of the stars on human affairs. We read there that

> The actual moment, in which human generation commences, is in fact, by nature, the moment of the conception itself; but, in efficacy with regard to subsequent events, it is the parturition, or birth. [. . .] In every case, however, where the actual time of conception may be ascertained, either casually or by observation, it is useful to remark the effective influence of the configuration of the stars as it existed at that time; and, from that influence, to infer the future personal peculiarities of mind and body.[. . .] But, if the time of conception cannot be precisely made out, that of the birth must be received as the original date of generation; for it is virtually the most important and is in no respect deficient, on comparison with the primary origin by conception, except in one view only; viz. that the origin by conception affords the inference of

> occurrences which take effect previously to the birth, whereas the origin by birth can, of course, be available only for such as arise subsequently.

Spending words on a pseudoscience like astrology may seem out of place here, but it is interesting to note that in this field, forgetting what Ptolemy had written, people ended up considering only the date of birth, even if the latter offers a good starting point to estimate the date of conception.

Or maybe not. First of all, there are often premature births. In this case too, we usually calculate the age of the individual from the day (s)he came to light, but, from a developmental biology point of view, this should not be considered equivalent to the date of a parturition at the end of a 9-month gestation. Yet many opportunities that the individual will have in his or her social life, for example the date on which (s)he will be of age, or whether (s)he may take part in a competition with age-restricted access, are based on this disputable administrative choice.

In other mammals, a choice between determining age from either fertilization or birth can be even more problematic. This is the case for species in which a prolonged arrest of embryonic development occurs. This phenomenon, technically described as embryonic diapause, is known for example in roe deer, where, around the middle of the nineteenth century, the acute observations of some hunters revealed a strange circumstance. Between the mating season and the time the females give birth to the young, the interval seems too long for the pregnancy of these animals. In fact, although mating takes place between July and August, no embryo is observed in the genital tract of roe deer females before the end of the year. This is due to greatly delayed implantation of the embryo on the uterine wall. Today, the phenomenon is known for more than 100 species of mammals, belonging to the most diverse groups – from seals to bears, from rodents to kangaroos. Examples are also found among reptiles and fishes. Of course, in oviparous animals the mechanism of embryonic diapause is completely different from what happens in mammals.

Temporary suspension of embryonic development is also known among invertebrates. This phenomenon has been studied, for example, in *Drosophila* and in *Caenorhabditis elegans*. In the latter, however, there may also be a digression, compared with the usual post-embryonic developmental schedule, which leads to a remarkable extension of the individual's life duration. This fact was discovered in a culture dish where environmental conditions had turned unfavourable for worms.

The body of nematodes is covered with an elastic but strong cuticle, and their growth is marked by moults, as in arthropods. Almost all nematodes, including *C. elegans*, reach the adult condition through four moults. The stages that precede the adult are often called larvae, but should more sensibly be called juveniles, because they do not differ much from the final stage, apart from not being sexually mature.

In 1937, the German nematode specialist A. G. Fuchs described, in some of these worms, a particular larval stage, on which serious attention only turned when it was rediscovered in *C. elegans*. In 1975 – when the worm had been a fashionable model species for only about 10 years – two researchers observed that, in an unfavourable environment, the moult at the end of the larva II stage did not produce a larva III, but a new type of larva, with mouth closed internally by a stopper and pharynx unable to suck. The animal can remain in this stage for a very long time: for this reason, it has been given the name of Dauer larva (from the German *Dauer*, duration). A transfer of these larvae to a fresh substrate rich in bacteria (their usual food) allows them to resume activity and continue development through the remaining larval stages up to the adult.

In *C. elegans*, an animal can remain in the Dauer larva stage up to 6 months, while the entire lifespan of the worm under normal conditions is 3 weeks. In this case, as in mammals with embryonic diapause, deciding the biological age of the animal can be problematic.

Different, but no less problematic, is the case of animals in which one half of the body ceases to develop at a precise time, while the other half continues to grow and change. This happens in *Hemioniscus balani*, a small crustacean parasitic of other crustaceans. This animal belongs to a very heterogeneous group of crustaceans called the isopods. Isopods include the woodlice, the most common and best known among the terrestrial crustaceans; other isopods live in fresh water, but it is in the sea that we find the greatest number and the greatest diversity of lifestyles among the representatives of this group. Several species are parasitic on fishes, others on crustaceans, as is the case of *H. balani*. This isopod species is a sequential hermaphrodite: it first goes through a male phase, then it becomes a female. This sex change does not involve the whole animal. Once the male phase is over, the front half of the body undergoes no further change and remains small, while the rear half continues moulting and becomes much more voluminous, allowing space for the eggs. In this case, periodization of the development valid for the whole body is not possible even based on moults – the most obvious, although not always satisfactory, frame of reference in arthropods, as discussed above.

The Mayfly and the Bristlecone Pine: Age and Senescence, if Any

For Each Species There Is Not Necessarily a Maximum Individual Age, and Death May Not Be Preceded by Senescence

Ephemeroptera is the name of the insect group in which mayflies are classified. The genus name *Ephemera* has had its place in the zoological classification since the publication, in 1758, of the tenth edition of *Systema naturae* by the Swedish naturalist Carolus Linnaeus, the work eventually chosen as the starting point of zoological nomenclature. *Ephemera* is a word of Greek origin meaning 'for one day only'. In fact, the winged phase of these insects lasts less than a day, preceded by a long series of aquatic stages. Even if the word 'programme' is not at all in accordance with my vision of the living, it is difficult to deny that a mayfly seems to be programmed to die immediately after its reproductive act.

An almost equally sudden conclusion of life is observed in other animals, usually less fragile than a mayfly, and also in some species of plants, some of which are robust and long-lived.

Tachygali versicolor, a native of the New World between Costa Rica and Colombia, is a tree species of the legume family which blooms only once in its life – we do not know at what age, but it must be remarkable, judging from the size of the plant, up to 40 metres high. The flowering time lasts a few weeks, then the plant ages quickly and dies within months. For this reason, the species is known locally as the suicide tree.

A rapid end of life after the only flowering season is also characteristic of most bamboo species. In some of these giant members of the grass family, the phenomenon is even more sensational, in that all individuals of the species, wherever in the world they grew up, bloom in the same year, then die (Figure 8.1).

Animals and plants with a long life, however, more often enjoy a reproductive period spread over several years, often followed in animals by a post-

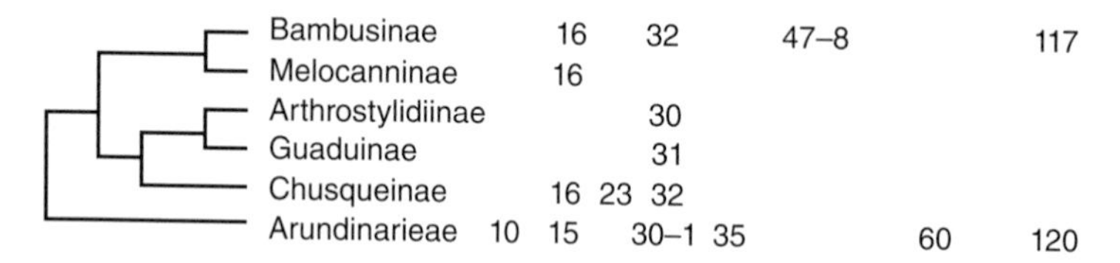

Figure 8.1 In most bamboo species, all individuals flower in the same year, and a fixed number of years separate one mass flowering from the previous one. These intervals are mostly of either 16 or 32 years (or close to that) but can be as high as 117 or 120 years. While remaining within this general scheme, there are differences between the main bamboo tribes; the tree on the left summarizes the evolutionary relationships among the latter.

reproductive season of accelerating senescence. Reasoning in terms of evolutionary biology, the post-reproductive decay of functions could be attributed to the impossibility, on the part of natural selection, of acting on any trait that the animal or plant exhibits when the reproductive season is over. The conditions of a senescent animal or plant do not affect its reproductive success, therefore we must expect a progressive accumulation of mutations with negative effects on the animal or plant in old age.

But this is not always the case. We know, as humans, that old individuals can have great importance among social animals: in this case, there is a positive selection in favour of individuals who age well, to the advantage of their younger relatives. In terms of developmental biology, however, other aspects are more interesting.

There are species in which senescence is unknown. If continuous growth can be taken as a sign of negligible senescence, good examples are the continuously growing animals mentioned in Chapter 1 (p. 7), such as groupers among fishes and elephants among mammals.

Most fungi, both unicellular yeasts and multicellular filamentous fungi, appear to be immortal. More pertinent, however, is the example of the freshwater planarian *Schmidtea mediterranea*, which constantly and seemingly permanently replaces the cells lost to physiological wear and tear with the progeny of neoblasts.

Dioscorea pyrenaica is a herbaceous plant whose individuals can live for more than 300 years. Precise observations on 260-year-old individuals revealed no drop in growth rate or reproductive capacity compared with younger specimens. In other plants, the life of an individual is short, but there are important examples of a resumption of developmental processes after reproduction. When a flower is fertilized and the complex events that lead to the production of seeds begin, the ovary undergoes transformations that end with the formation of the fruit. The non-reproductive parts of the flower generally dry out or fall off. However, in some species the calyx resumes developing at the end of flowering. A lengthening of the sepals, for example, is observed in the myosotis and in various species of the mint family. But these are very modest changes, compared with what happens in *Physalis*. In this plant, the calyx grows a lot and forms a sort of balloon enclosing the fruit: this is the 'Chinese lantern' with its striking orange-red colour.

In any case, senescence appears to be a likely fate for the post-reproductive segment of the life of a multicellular organism. But what happens to unicellulars? Is there also a change in physiological conditions here comparable to the senescence of an old animal?

The answer is yes. To clarify facts, let us resort once more to the ciliates. In Chapter 2 (p. 31) I mentioned the separation that exists in these unicellular organisms between reproduction and sexuality. The latter, as said, consists in a renewal of the genome through the exchange of nuclei between two partners (conjugants). An ex-conjugant just after this exchange is therefore a suitable individual on which to begin observations. Under favourable environmental conditions, this individual undergoes cell division and eventually gives rise to a clone of increasing size. However, this process has a limit: in the ciliates of the genus *Tetrahymena*, for example, this varies, depending on the strains, between 40 and 1500 divisions. In the initial phase of this numerical progression, individual ciliates cannot perform conjugation, but this occurs with increasing frequency in older clones. In ciliates, the sexual process is a mechanism allowing erasure of the effects of ageing. Individuals that do not practice conjugation may continue to multiply, but will enter a period of gradual senescence through which the rate of cell division gradually slows down and eventually leads to the extinction of the clone.

Senescence, however, does not seem to be universal, even among the ciliates. Strains of *Tetrahymena pyriformis* have been cultured for more than 50 years without performing conjugation, yet failed to show any signs of senescence.

Concluding Remarks

On 23 October 1908, William Bateson, the first scientist to use the term genetics to describe the science of heredity, delivered his inaugural lecture as Professor of Biology at the University of Cambridge. The text of this official talk contains a sentence worth transcribing here:

> if I may throw out a word of counsel to beginners, it is: Treasure your exceptions! [. . .] Keep them always uncovered and in sight. Exceptions are like the rough brickwork of a growing building which tells that there is more to come and shows where the next construction is to be.

I could not leave my readers with more fitting words. In 1894, Bateson had published a volume titled *Materials for the Study of Variation*. It was mainly intended as a contribution to evolutionary biology, as revealed by the subtitle: *Treated with Especial Regard to the Discontinuity in the Origin of Species*. Variation, however, deserves to be studied in all provinces of the life sciences, especially those in which too strict a focus on a bunch of model species is the premise for incautious haste to proclaim general principles. As we have seen in this book, this is often the case with developmental biology.

This is why, on this last page, I am happy to repeat Bateson's counsel to treasure our exceptions. . . except for a footnote: whenever a rule suffers exceptions, the rule is arguably a different one.

Summary of Common Misunderstandings

Common misunderstandings about development, and responses stemming from this book:

To study development is to study multicellularity. This is a traditional but unjustified restriction of developmental biology. Structural (not simply metabolic) changes are also observed in single-celled organisms (very conspicuous, for example, in trypanosomes) and in the production of the single-cell phase (such as eggs and sperm cells) in the biological cycle.

There are significant similarities between the early developmental stages of animals and plants. Animals and plants are the two most successful groups among those that have evolved multicellularity, but they derive from different single-celled ancestors. The organization of the multicellular phase of their biological cycle is based on different cellular properties and is produced through very different developmental pathways. In animals, at the end of embryonic development the entire structure of the organism is almost always delineated. In flowering plants, the seedling formed during the so-called embryonic development contains only the shoot with the first leaves and the radicle, while the entire structure of the plant, including almost all the leaves and all the flowers, will form from meristems (groups of stem cells) generated through the entire life of the plant.

Development is a sustainably adaptive process. This is often not true. The growth of a tumour, for example, is a developmental process, albeit tragically maladaptive for the organism from which the mass of cancerous cells originates.

Development is the history of an individual from egg to adult. An individual's development from egg to adult *is* a story of development, but development is not only that. In cases of asexual reproduction, the new individual is formed not from an egg but from a bud or a simple fragment of the parent. In cellular slime moulds, a multicellular phase originates by aggregation of isolated cells. These two examples counter another common opinion, that at the beginning of an individual's development there must necessarily be a single-celled bottleneck. In some organisms, such as the colonial sea squirt *Botryllus*, morphologically similar individuals generated sexually or asexually develop through very different series of stages.

The adult stage is the target of development. Adultocentrism – that is, the notion that the adult stage is the target of development – is unacceptable, both because it implicitly gives development a purpose, a notion foreign to science, and because it does not apply to the biology of many organisms. In the common use of the term adult, two different notions are confused: adult as reproductively mature stage and as a stage that maintains its morphological organization until the onset of senescence or death. However, reproductive maturity and the presence of definitive morphological condition are not always associated. As to the widespread opinion that life beyond the end of reproductive age is uninteresting, not all organisms experience senescence; at any rate, even changes associated with ageing can be considered as aspects of developmental biology.

The egg is the least specialized cell. This notion is linked to the idea that development is the history of the individual from egg to adult. In fact, the egg is one of the most specialized cells, and the process through which an egg is produced (oogenesis) is per se a developmental process.

Development is the actualization of a genetic programme. The events that occur during any developmental phase or process are characterized by variations in the expression of a number of genes that often amount to an important percentage of all protein-coding genes. However, in the genome there are not specific genes for specific phenotypic traits or genes responsible for the production of a major organ such as heart or eye, and there would be no gene expression were it not for the whole machinery of the cell. Furthermore, all developmental processes take place in a specific environmental context that influences the expression of genes; sometimes, environmental differences can cause individuals with the same genome to develop into very different phenotypes, for example male and female.

Regeneration is one of the great mysteries of biology. There is no reason to consider regeneration as the expression of a hidden finalism intrinsic to living matter. Instead, regeneration is the reactivation – induced by injury or by physiological loss of a body part, as in the annual fall of deer antlers – of a proliferative and morphogenetic activity that is broadly comparable to embryonic development or to asexual reproduction by fragmentation. Furthermore, regeneration is not necessarily adaptive.

Classification

All genera and species as well as higher (suprageneric) taxa mentioned in the book are distributed here according to the current classification.

Bacteria: Bacteroidetes: *Algoriphagus machipongonensis*; Cyanobacteria; Firmicutes; Proteobacteria: *Aliivibrio fischeri*, *Buchnera*, *Caulobacter vibrioides*, *Wolbachia*

Mycetozoa (slime moulds): cellular slime mould (*Dictyostelium discoideum*), *Physarum*

Fungi

Microsporidia (unicellular, parasitic on animals): *Schroedera airthreyi*

Ascomycota (ascomycetes: yeasts, many moulds, most lichens, truffles and many parasitic, generally microscopic fungi): baker's yeast (*Saccharomyces cerevisiae*), *Candida albicans*, lichens incl. *Aspicilia californica*, red bread mould (*Neurospora crassa*), *Neotyphodium*, *Epichloë*

Basidiomycota (basidiomycetes: most mushrooms, a few lichens, some parasitic fungi): *Coprinopsis cinerea*

Choanozoa (choanoflagellates): *Salpingoeca rosetta*

Metazoa (animals)

Porifera (sponges) (glass) hexactinellids

Ctenophora (comb jellies): *Mertensia ovum*

Cnidaria (polyps and medusae)

Anthozoa (polyps, with skeleton – corals, madrepores: *Acropora sarmentosa* – or without: sea anemones)

Hydrozoa (polyps only: *Hydra vulgaris, H. oligactis*; polyp and medusa alternating in the life cycle: *Clytia gregaria, Hydractinia echinata, Turritopsis dohrnii*)
Cubozoa (polyp metamorphosing into medusa)
Scyphozoa (mostly polyp and medusa alternating in the life cycle): *Chrysaora isoscella*
Gastrotricha (gastrotrichs)
Syndermata (rotifers, mostly free-living, and the parasitic acanthocephalans): monogononts
Rhabditophora (flatworms, incl. free-living forms (*Macrostomum lignano* and planarians: *Dendrocoelum lacteum, Dugesia subtentaculata, Schmidtea mediterranea*) and parasitic worms (Cestoda or tapeworms: *Taenia solium, Hymenolepis diminuta*; Digenea (flukes): *Diplozoon*; Monogenea: *Wedlia bipartita, Gyrodactylus*))
Orthonectida (orthonectids)
Nemertini (nemertines or ribbon worms): *Lineus*
Mollusca (molluscs): Bivalvia (clams, oysters, scallops): oyster, *Tridacna*; Gastropoda (snails and slugs): *Lottia gigantea*, abalone (*Haliotis asinina*), *Turbinella pyrum*, Clausiliidae incl. *Balaea perversa, Radix peregra, Lymnaea stagnatilis, Biomphalaria glabrata, Elysia chlorotica*; Cephalopoda (octopuses and squids): bobtail squid (*Euprymna scolopes*), octopus
Annelida (annelids): *Ramisyllis multicaudata, Sphaerosyllis hystrix, Syllis gracilis, S. ramosa*, earthworms, leeches
Bryozoa (bryozoans): *Plumatella*
Loricifera (loriciferans)
Nematoda (nematodes or roundworms): *Brugia malayi, Caenorhabditis elegans*
Tardigrada (water bears)
Arthropoda (arthropods)
Arachnida (arachnids)
Acari (mites)
Myriapoda (myriapods)
Chilopoda (centipedes): Scutigeromorpha (house centipedes); Lithobiomorpha (brown centipedes); Scolopendromorpha (scolopenders); Geophilomorpha
Diplopoda (millipedes)

Pancrustacea (crustaceans and insects)
Rhizocephala (a group of parasitic crustaceans): *Sacculina carcini*
Isopoda (isopods): *Hemioniscus balani*, woodlice
Decapoda (shrimps, lobsters, crabs)
Hexapoda (insects)
Ephemeroptera (mayflies)
Blattodea (cockroaches and termites)
Hemiptera (plant bugs, aphids, cicadas): *Magicicada*, Coreidae
Coleoptera (beetles): Dermestidae (skin beetles): *Trogoderma glabrum*; Coccinellidae (ladybirds); Cerambycidae (longhorn beetles): *Aromia moschata*; Chrysomelidae (leaf beetles): *Callosobruchus chinensis*
Diptera (flies, midges, mosquitoes): Chironomidae (non-biting midges): *Chironomus*, *Paratanytarsus grimmi*; Cecidomyidae (gall midges); Culicidae (mosquitoes): *Aedes aegypti*; Drosophilidae: *Drosophila melanogaster*
Lepidoptera (butterflies and moths): large white (*Pieris brassicae*), *Nemoria aizonaria*
Hymenoptera (bees, wasps and ants): *Ageniaspis fuscicollis*, *Asobara*, *Copidosoma truncatellum*
Echinodermata (echinoderms)
Asteroidea (sea stars)
Holothuroidea (sea cucumbers): *Apostichopus japonicus*
Echinoidea (sea urchins): *Heliocidaris erythrogramma*, *H. tuberculata*
Chordata (chordates)
Tunicata (tunicates)
Ascidiacea (sea squirts): *Botryllus schlosseri*
Appendicularia (appendicularians)
Vertebrata (vertebrates)
Chondrichthyes (sharks, skates, rays): bull shark (*Carcharias taurus*), cigar shark (*Isistius brasiliensis*)
Actinopterygii (ray-finned fishes): swordfish, zebrafish (*Danio rerio*), *Trimma okinawae*, grouper, annual fish (*Cynolebias*), sea devils (Ceratioidei: *Ceratias holboellii*, *Photocorynus spiniceps*), monkfish (*Lophius piscatorius*)

Stegocephalia (all extinct): *Acanthostega, Ichthyostega, Tulerpeton*

Amphibia (amphibians): Urodela (salamanders and newts): fire salamander (*Salamandra salamandra*), newt; Anura (frogs and toads): frog

Reptilia (reptiles and birds): Lepidosauria (lizards and snakes): pythons, rattlesnakes; Testudinata (tortoises, terrapins and turtles): loggerhead sea turtle (*Caretta caretta*), giant tortoises of the Galapagos; Crocodylia (crocodiles, alligators): American alligator; Aves (birds): chicken

Mammalia (mammals): Monotremata (echidnas and platypus): platypus; Dasyuromorphia: (carnivorous Australian marsupials): Tasmanian devil; Diprotodontia (mostly herbivorous Australian marsupials): kangaroos; Cingulata: nine-banded armadillo; Rodentia (rodents): house mouse (*Mus musculus*); Primates (man, apes, monkeys): chimpanzee, man; Cetartiodactyla (heterogeneous group incl. whales, ruminants and others): red deer, roe deer, bison, giraffes; Proboscidata (elephants); Carnivora (carnivores): bears, seals

Plantae (plants)

Chlorophyta (main group of green algae): *Acetabularia*

Charophyta (also green algae): *Spirogyra*

Bryophyta (mosses): *Aphanorrhegma patens*

Polypodiopsida (ferns): cinnamon fern (*Osmundastrum cinnamomeum*)

Spermatopsida (seed plants)

Gymnospermae (naked-seed plants): *Welwitschia mirabilis*; Pinales (conifers): Great Basin bristlecone pine (*Pinus longaeva*), giant redwood (*Sequoiadendron giganteum*)

Angiospermae (flowering plants): Araceae (arum family): *Monstera*; Dioscoreaceae (yam family): *Dioscorea pyrenaica*; Asparagaceae (asparagus family): hyacinth; Poaceae (grass family): bamboo, rice (*Oryza sativa*); Ranunculaceae (buttercup family): lesser spearwort (*Ranunculus flammea*); Fabaceae (pea family): *Anthyllis cytisoides*, black locust (*Robinia pseudacacia*), suicide tree (*Tachygali versicolor*); Cannabaceae (hemp family): hop (*Humulus lupulus*); Betulaceae (birch family): hazel

(*Corylus*); Meliaceae (mahogany family): *Guarea, Chisocheton*; Malvaceae (mallow family): lime (*Tilia*); Brassicaceae (mustard family): marsh yellow cress (*Rorippa palustris*), *Arabidopsis lyrata,* thale cress (*Arabidopsis thaliana*), cauliflower; Polygonaceae (buckwheat family): Alpine bistort (*Bistorta vivipara*); Boraginaceae (borage family): *Myosotis*; Convolvulaceae (morning glory family): bindweed (*Convolvulus*); Solanaceae (potato family): Chinese lantern (*Physalis*), tomato (*Solanum lycopersicum*); Oleaceae (olive family): ash (*Fraxinus*); Plantaginaceae (plantain family): snapdragon (*Antirrhinum majus*), *Littorella*; Lamiaceae (mint family); Asteraceae (aster family): corn marigold (*Glebionis segetum*); Adoxaceae (elderberry family): elder (*Sambucus*); Caprifoliaceae (honeysuckle family): wild teasel (*Dipsacus fullonum*), honeysuckle (*Lonicera*)

Euglenozoa (a group of unicells incl. both parasites and free-living species): *Trypanosoma brucei*

Xanthophyceae (yellow-green algae): *Vaucheria litorea*

Dinoflagellata (a group of flagellated, mostly photosynthetic unicells): *Symbiodinium*

Ciliata (ciliates): *Paramecium, Stentor coeruleus, Tetrahymena pyriformis*

References and Further Reading

Chapter 1

Textbooks on developmental biology: Wolpert, L., Tickle, C., and Martinez Arias, A. (2015). *Principles of Development*, 5th edn. Oxford: Oxford University Press; Gilbert, S. F., and Barresi, J. F. (2016). *Developmental Biology*, 11th edn. Sunderland, MA: Sinauer.

On a possible definition of development: Minelli, A. (2011). Development, an open-ended segment of life. *Biological Theory* 6: 4–15; Pradeu, T., Laplane, L., Prévot, K., et al. (2016). Defining "Development". *Current Topics in Developmental Biology* 117: 171–183.

On a theory of development: Minelli, A., and Pradeu, T. (eds.) (2014). *Towards a Theory of Development*. Oxford: Oxford University Press.

The quotation on p. 2 is from Bernard, C. (1878). *Leçons sur les phénomènes de la vie communs aux animaux et aux végétaux*. Volume 1. Baillière, Paris (pp. 331–333, my translation).

On adultocentrism: Minelli, A. (2003). *The Development of Animal Form*. Cambridge: Cambridge University Press. Davidson's sentences quoted on p. 6 are from Davidson, E. H. (1991). Spatial mechanisms of gene regulation in metazoan embryos. *Development* 113: 1–26.

On choanoflagellates: Hoffmeyer, T. T., and Burkhardt, P. (2016). Choanoflagellate models – *Monosiga brevicollis* and *Salpingoeca rosetta*. *Current Opinion in Genetics and Development* 39: 42–47.

On morphological and functional changes in python anatomy during and after feeding: Andersen, J. B., Rourke, B. C., Caiozzo, V. J., Bennett, A. F., and

Hiàcks, J. W. (2005). Postprandial cardiac hypertrophy in pythons. *Nature* 434: 37–38; Andrew, A. L., Card, D. C., Ruggiero, R. P., et al. (2015). Rapid changes in gene expression direct rapid shifts in intestinal form and function in the Burmese python after feeding. *Physiological Genomics* 47: 147–157.

On *Elysia chlorotica*: Sultan, S. (2015). *Organism and Environment: Ecological Development, Niche Construction, and Adaptation*. New York: Oxford University Press.

On the history of embryology: Needham, J. (1959). *A History of Embryology*, 2nd edn. Cambridge: Cambridge University Press; Gilbert, S. F. (ed.) (1991). *A Conceptual History of Modern Embryology*. New York: Plenum.

On the history of research on model organisms (*Drosophila, Caenorhabditis elegans*): Kohler, R. E. (1994). Lords of the Fly*: Drosophila* Genetics and the Experimental Life. Chicago: University of Chicago Press; de Chadarevian, S. (1998). Of worms and programs: *Caenorhabditis elegans* and the study of development. *Studies in History and Philosophy of the Biological and Biomedical Sciences* 29: 81–105; Minelli, A., and Baedke, J. (2014). Model organisms in evo-devo: Promises and pitfalls of the comparative approach. *History and Philosophy of the Life Sciences* 36: 42–59.

On the ABC model of flower parts specification: Coen, E. S., and Meyerowitz, E. M. (1991). The war of the whorls: Genetic interactions controlling flower development. *Nature* 353: 31–37; Bowman, J. L., Smyth, D. R., and Meyerowitz, E. M. (1991). Genetic interactions among floral homeotic genes of *Arabidopsis*. *Development* 112: 1–20.

Chapter 2

Papers cited in the chapter's opening paragraphs: Chandebois, R. (1977). Cell sociology and the problem of position effect: pattern formation, origin and role of gradients. *Acta Biotheoretica* 26: 203–238; Wolpert, L., Ghysen, A., and García-Bellido, A. (1998). Debatable issues. *International Journal of Developmental Biology* 42: 511–518 (L. Wolpert, p. 515).

On Loricifera: Kristensen, R. M. (2002). An introduction to Loricifera, Cycliophora, and Micrognathozoa. *Integrative and Comparative Biology* 42: 641–651.

On the *Drosophila chico* mutant: Böhni, R., Riesgo-Escovar, J., Oldham, S., et al. (1999). Autonomous control of cell and organ size by CHICO, a *Drosophila* homolog of vertebrate IRS1–4. *Cell* 97: 865–876.

On the transcription of very large genes in early embryonic development: O'Farrell, P. H. (1992). Big genes and little genes and deadlines for transcription. *Nature* 359: 366–367.

On the hands and feet of *Ichthyostega* and *Acanthostega*: Coates, M. I., and Clack, J. A. (1990). Polydactyly and the earliest known tetrapod limbs. *Nature* 347: 66–69.

On developmental biology as the study of multicellularity: Bonner, J. T. (2001). *First Signals: The Evolution of Multicellular Development*. Princeton: Princeton University Press.

On *Trypanosoma*: Matthews, K. R. (2005). The developmental cell biology of *Trypanosoma brucei*. *Journal of Cell Science* 118: 283–290.

On *Acetabularia*: Hämmerling, J. (1953). Nucleo-cytoplasmic relationships in the development of *Acetabularia*. *International Review of Cytology* 2: 475–498; Berger, S., and Liddle, L. B. (2003). The life cycle of *Acetabularia* (Dasycladales, Chlorophyta): Textbook accounts are wrong. *Phycologia* 42: 204–207.

On ciliates: Tartar, V. (1961). The Biology of *Stentor*. Oxford: Pergamon Press.

On anucleate animal cells: Polilov, A. A. (2012). The smallest insects evolve anucleate neurons. *Arthropod Structure and Development* 41: 27–32.

On embryonic development in the fruit fly, including the syncytial phase: Lawrence, P. A. (1992). *The Making of a Fly*. Oxford: Blackwell.

On syncytia in glass sponges: Leys, S. P. (2003). The significance of syncytial tissues for the position of the Hexactinellida in the Metazoa. *Integrative and Comparative Biology* 43: 19–27.

On the cytoplasmic strands connecting plant cells: Lucas, W. J., Ding, B., and Van der Schoot, C. (1993). Plasmodesmata and the supracellular nature of plants. *New Phytologist* 125: 435–476.

Chapter 3

On cellular slime moulds (*Dictyostelium discoideum*): Bonner, J. T. (1959). *The Cellular Slime Molds*. Princeton: Princeton University Press; Loomis, W. F. (ed.) (1982). *The Development of* Dictyostelium discoideum. New York: Academic Press.

On biological individuality: Buss, L. (1987). *The Evolution of Individuality*. Princeton: Princeton University Press; Santelices, B. (1999). How many kinds of individual are there? *Trends in Ecology and Evolution* 14: 152–155; Wilson, J. (1999). *Biological Individuality: The Identity and Persistence of Living Entities*. Cambridge: Cambridge University Press; Godfrey-Smith, P. (2009). *Darwinian Populations and Natural Selection*. Oxford University Press: New York; Bouchard, F., and Huneman, P. (eds.) (2013). *From Groups to Individuals. Evolution and Emerging Individuality*. Cambridge, MA: The MIT Press; Pradeu, T. (2016). Organisms or biological individuals? Combining physiological and evolutionary individuality. *Biology and Philosophy* 31: 797–817; Fields, C., and Levin, M. (2018). Are planaria individuals? What regenerative biology is telling us about the nature of multicellularity. *Evolutionary Biology* 45: 237–247.

On genet and ramet: Harper, J. L., and White, J. (1974). The demography of plants. *Annual Review of Ecology and Systematics* 5: 419–463.

On chimerism and mosaicism in stony corals: Schweinsberg, M., Weiss, L. C., Striewski, S., Tollrian, R., Lampert, K. P. (2015). More than one genotype: how common is intracolonial genetic variability in scleractinian corals? *Molecular Ecology* 24: 2673–2685.

On mutation rates in humans: Kong, A., Frigge, M. L., Masson, G., et al. (2012). Rate of de novo mutations and the importance of father's age to disease risk. *Nature* 488: 471–475; Sun, J. X., Helgason, A., Masson, G., et al. (2012). A direct characterization of human mutation based on microsatellites. *Nature Genetics* 44: 1161–1165.

On polyembryony: Craig, S. F., Slobodkin, L. B., Wray, G. A., and Biermann, C. H. (1997). The 'paradox' of polyembryony: A review of the cases and a hypothesis for its evolution. *Evolutionary Ecology* 11: 127–143; Gordon, S. D., and Strand, M. R. (2009). The polyembryonic wasp *Copidosoma floridanum* produces two castes by differentially parceling the germ line during embryo proliferation. *Development, Genes and Evolution* 219: 445–454.

On developmental scaffolds: Minelli, A. (2016). Scaffolded biology. *Theory in Biosciences* 135: 163–173.

On animals of Greek mythology revisited from the perspective of comparative morphology and morphogenetics: Minelli, A. (2015). Constraints on animal (and plant) form in nature and art. *Art and Perception* 3: 265–281.

On chimeric pairs and immune problems in angler fishes: Regan, C. T. (1925). Dwarfed males parasitic on the females of oceanic angler-fishes (Pediculati, Ceratioidea). *Proceedings of the Royal Society of London B* 97: 386–400; Dubin, A., Jørgensen, T. E., Moum, T., Johansen, S. D., and Jakt, L. M. (2019).

Complete loss of the MHC II pathway in an anglerfish, *Lophius piscatorius*. *Biology Letters* 15: 20190594.

On *Sacculina carcini*: Høeg, J. T. (1987). Male cypris metamorphosis and a new male larval form, the trichogon, in the parasitic barnacle *Sacculina carcini* (Crustacea: Cirripedia: Rhizocephala). *Philosophical Transactions of the Royal Society of London* 317B: 47–63.

On lichens and plant galls: Sanders, W. (2006). A feeling for the superorganism: Expression of plant form in the lichen thallus. *Botanical Journal of the Linnean Society* 150: 89–99; Minelli, A. (2017). Lichens and galls – two families of chimeras in the space of form. *Azafea* 19: 91–105.

Chapter 4

On inertial models in biology: Gayon, J. (1998). *Darwinism's Struggle for Survival*. Cambridge: Cambridge University Press; Minelli, A. (2011). A principle of developmental inertia. In Hallgrímsson, B. and Hall, B. K. (eds.) *Epigenetics: Linking Genotype and Phenotype in Development and Evolution*. Berkeley–Los Angeles–London: University of California Press, pp. 116–133.

Darwin's quotations about monsters and deformities are from Barrett, P. H., Gautrey, P. J., Herbert, S., Kohn, D. and Smith, S. (eds.) (1987). *Charles Darwin's Notebooks, 1836–1844*. Cambridge: Cambridge University Press (Notebook B, p. 199; Notebook C, p. 259).

On the intertwining of genes, environment and epigenetic factors in development: Oyama, S. (2000). *The Ontogeny of Information*, 2nd edn. Durham, NC: Duke University Press; Kupiec, J. J. (2009). *The Origin of Individuals*. Singapore: World Scientific; Griesemer, J. (2019). Towards a theory of extended development. In Fusco, G. (ed.) *Perspectives on Evolutionary and Developmental Biology*. Padova: Padova University Press, pp. 319–334.

On extra-embryonic membranes and other annexes: Gilbert, S. F., and Raunio, A. M. (1997). *Embryology: Constructing the Organism*. Sunderland, MA: Sinauer; Panfilio, K. A. (2008). Extraembryonic development in insects and the acrobatics of blastokinesis. *Developmental Biology* 313: 471–491.

On Roux and Driesch in the context of the history of embryology since the last decades of the nineteenth century: Gilbert, S. F. (ed.) (1991). *A Conceptual History of Modern Embryology*. New York: Plenum.

On the reaggregation of dissociated cells in sponges: Wilson, H. V. (1907). On some phenomena of coalescence and regeneration in sponges. *Journal of Experimental Zoology* 5: 245–258; Custodio, M. R., Prokic, I., Steffen, R., et al. (1998). Primmorphs generated from dissociated cells of the sponge *Suberites domuncula*: A model system for studies of cell proliferation and cell death. *Mechanisms of Ageing and Development* 105: 45–59; Lavrov, A. I., and Kosevich, I. A. (2016). Sponge cell reaggregation: Cellular structure and morphogenetic potencies of multicellular aggregates. *Journal of Experimental Zoology* 325A: 158–177.

On the cell lineage of *Caenorhabditis elegans*: Packer, J. S., Zhu, Q., Huynh, C., et al. (2019). A lineage-resolved molecular atlas of *C. elegans* embryogenesis at single cell resolution. *Science* 365, eaax1971.

On the medusa-to-polyp transition in *Turritopsis*: Bavestrello, G., Sommer, C., Sarà, M., and Hughes, R. G. (1992). Bi-directional conversion in *Turritopsis nutricula*. In Bouillon, J., Boero, F., Cicogna, F., Gili, J. M., and Hughes, R. G. (eds.), *Aspects of Hydrozoan Biology*. *Scientia Marina* 56: 137–140; Piraino, S., Boero, F., Aeschbach, B., and Schmid, V. (1996). Reversing the life cycle: medusae transforming into polyps and cell transdifferentiation in *Turritopsis nutricula* (Cnidaria, Hydrozoa). *Biological Bulletin* 190: 302–312.

On Lazarus developmental features: Minelli, A. (2003). *The Development of Animal Form*, cit.

On organogenesis as resulting from interwined, non-organ-specific mechanisms: Carmeliet, P., and Tessier-Lavigne, M. (2005). Common mechanisms of nerve and blood vessel wiring. *Nature* 436: 193–200; Dunwoodie, S. L. (2007). Combinatorial signalling in the heart orchestrates cardiac induction, lineage specification and chamber formation. *Seminars in Cell and Developmental Biology* 18: 54–66; Tao, Y. and Schulz, R. A. (2007). Heart development in *Drosophila*. *Seminars in Cell and Developmental Biology* 18: 3–15; Minelli, A. (2009). *Perspectives in Animal Phylogeny and Evolution*. Oxford: Oxford University Press.

On the development of the pilidium larva of nemertines: Maslakova, S. A. (2010). Development to metamorphosis of the nemertean pilidium larva. *Frontiers in Zoology* 7: 30.

Von Baer's law of development (p. 70) is found on p. 224 of von Baer, K. E. (1828). *Über Entwickelungsgeschichte der Thiere: Beobachtung und Reflexion*. Volume 1. Königsberg: Bornträger. Extensive discussion in Gould, S. J. (1977). *Ontogeny and Phylogeny*. Cambridge, MA: Belknap Press of Harvard University Press.

On phylotypic stage and zootype: Sander, K. (1983). The evolution of patterning mechanisms: gleanings from insect embryogenesis and spermatogenesis. In Goodwin, B. C., Holder, N., and Wylie, C. C. (eds.) *Development and Evolution*. Cambridge: Cambridge University Press, pp. 124–137; Slack, J. M. W., Holland, P. W. H., and Graham, C. F. (1993). The zootype and the phylotypic stage. *Nature* 361: 490–492; Duboule, D. (1994). Temporal colinearity and the phylotypic progression: A basis for the stability of a vertebrate Bauplan and the evolution of morphologies through heterochrony. *Development* (Supplement): 135–142; Richardson, M. K. (1995). Heterochrony and the phylotypic period. *Developmental Biology* 172: 412–421; Wu, L., Ferger, K. E., and Lambert, J. D. (2019). Gene expression does not support the developmental hourglass model in three animals with spiralian development. *Molecular Biology and Evolution* 36: 1373–1383.

On the embryonic and larval development of *Heliocidaris tuberculata* versus *H. erythrogramma*: Wray, G. A., and Raff, R. A. (1991). The evolution of developmental strategy in marine invertebrates. *Trends in Ecology and Evolution* 6: 45–50.

Chapter 5

On the embryonic versus blastogenetic development in *Botryllus schlosseri*: Manni, L., and Burighel, P. (2006). Common and divergent pathways in alternative developmental processes of ascidians. *BioEssays* 28: 902–912; Alié, A., Hiebert, L., Scelzo, M., and Tiozzo, S. (2020). The eventful history of non-embryonic development in colonial Tunicates. *Journal of Experimental Zoology B Molecular and Developmental Evolution*. DOI: 10.1002/jez.b.22940.

On larval reproduction in the jellycomb *Mertensia ovum*: Jaspers, C., Haraldsson, M., Bolte, S., Reusch, T. B. H., Thygesen, U. H., and Kiørboe, T. (2012). Ctenophore population recruits entirely through larval reproduction in the central Baltic Sea. *Biology Letters* 8: 809–812.

On the history of preformism/pre-existence and epigenesis: Bowler, P. J. (1971). Preformation and pre-existence in the seventeenth century. *Journal of the History of Biology* 4: 221–244; Roe, S. A. (1981). *Matter, Life and Generation*. Cambridge: Cambridge University Press; Pinto-Correia, C. (1997). *The Ovary of Eve*. Chicago: University of Chicago Press; Cobb, M. (2006). *The Egg and Sperm Race*. Bloomsbury: The Free Press; Pyle, A. (2006). Malebranche on animal generation. Preexistence and the microscope. In J. E. H. Smith (ed.) *The Problem of Animal Generation in Early Modern Philosophy*. Cambridge: Cambridge University Press, pp. 194–214.

On the boxed generations of *Gyrodactylus*: Cable, J., and Harris, P. D. (2002). Gyrodactylid developmental biology: Historical review, current status, and future trends. *International Journal of Parasitology* 32: 255–280.

On recent proposals about the definition of generation: Gorelik, R. (2012). Mitosis circumscribes individuals; sex creates new individuals. *Biology and Philosophy* 27: 871–890; Minelli, A. (2014). Developmental disparity. In Minelli, A. and Pradeu, T. (eds.) *Towards a Theory of Development*. Oxford: Oxford University Press, pp. 227–245.

On planarian regeneration: Reddien, P. W., and Sánchez Alvarado, A. (2004). Fundamentals of planarian regeneration. *Annual Review of Cell and Developmental Biology* 20: 725–757; Egger, B., Ladurner, P., Nimeth, K., Gschwentner, R., and Rieger, R. (2006). The regeneration capacity of the flatworm *Macrostomum lignano* – on repeated regeneration, rejuvenation, and the minimal size needed for regeneration. *Development Genes and Evolution* 216: 565–577; Sánchez Alvarado, A. (2012). What is regeneration, and why look to planarians for answers? *BMC Biology* 10: 88; Rink, J. C. (2013). Stem cell systems and regeneration in planaria. *Development Genes and Evolution* 223: 67–84; Reddien, P. W. (2018). The cellular and molecular basis for planarian regeneration. *Cell* 175: 327–345.

On hydra regeneration: Bode, H. R., Berking, S., David, C. N., et al. (1973). Quantitative analysis of cell types during growth and morphogenesis in hydra. *Wilhelm Roux's Archives of Developmental Biology* 171: 269–285; Shimizu, H., Sawada, Y., and Sugiyama, T. (1993) Minimum tissue size required for hydra regeneration. *Developmental Biology* 155: 287–296; Martinez, D. E., and Bridge, D. (2012). *Hydra*, the everlasting embryo, confronts aging. *International Journal of Developmental Biology* 56: 479–487.

On morphallaxis: Morgan, T. H. (1901). *Regeneration*. New York: Macmillan.

On reproduction and development in monogonont rotifers: Gilbert, J. J. (2003). Environmental and endogenous control of sexuality in a rotifer life cycle: Developmental and population biology. *Evolution and Development* 5: 19–24.

On the transmissible Tasmanian devil tumour: Pearse, A. M., and Swift, K. (2006). Allograft theory: Transmission of devil facial-tumour disease. *Nature* 439: 549; Weiss, R. A. (2018). Open questions: Knowing who's who in multicellular animals is not always as simple as we imagine. *BMC Biology* 16: 115.

Chapter 6

The classic papers revisiting in molecular terms É. Geoffroy Saint-Hilaire's idea of inversion of the dorso-ventral axis between arthropods and vertebrates: Arendt, D., and Nübler-Jung, K. (1994). Inversion of dorsoventral axis? *Nature* 371: 26; De Robertis, E. M., and Sasai, Y. (1996). A common plan for dorsoventral patterning in Bilateria. *Nature* 380: 37–40.

Critical views of the current gene programme paradigm: Nijhout, H. F. (1990). Metaphors and the role of genes in development. *BioEssays* 12: 441–446; Keller, E. (2002). *The Century of the Gene*. Cambridge, MA: Harvard University Press.

On the multiple products of alternative splicing: Wojtowicz, W. M., Flanagan, J. J., Millard, S. S., Zipursky, S. L., and Clemens, J. C. (2004). Alternative splicing of *Drosophila* Dscam generates axon guidance receptors that exhibit isoform-specific homophilic binding. *Cell* 118: 619–633.

On the embryo as a computable system: Wolpert, L. (1994). Do we understand development? *Science* 266: 571–572; Rosenberg, A. (1997). Reductionism redux: computing the embryo. *Biology and Philosophy* 12: 445–470; Laublicher, M. S., and Wagner, G. P. (2001). How molecular is molecular developmental biology? A reply to Alex Rosenberg's reductionism redux: computing the embryo. *Biology and Philosophy* 16: 53–68.

On combined genetic and mechanical control of morphogenesis: Chen, Q., Jiang, L., Li, C., et al. (2012). Haemodynamics-driven developmental pruning of brain vasculature in zebrafish. *PLoS Biology* 10: e1001374.

On master control genes: Gehring, W. J. (1998). *Master Control Genes in Development and Evolution*. New Haven: Yale University Press.

On alternative views on the role of genes in development: Akam, M. (1998). Hox genes: from master genes to micromanagers. *Current Biology* 8: R676–R678; Davidson, E. H., Rast, J. P., Oliveri, P., et al. (2002). A genomic regulatory network for development. *Science* 295: 1669–1678; Davidson, E. H. (2006). *The Regulatory Genome: Gene Regulatory Networks in Development and Evolution*. San Diego: Academic Press.

On gene expression throughout the whole development: Arbeitman, M. N., Furlong, E. E. M., Imam, F., et al. (2002). Gene expression during the life cycle of *Drosophila melanogaster*. *Science* 297: 2270–2275; Koutsos, A. C., Blass, C.,

Meister, S. et al. (2007). Life cycle transcriptome of the malaria mosquito *Anopheles gambiae* and comparison with the fruitfly *Drosophila melanogaster*. *Proceedings of the National Academy of Sciences of the United States of America* 104: 11304–11309; Graveley, B., Brooks, A., Carlson, J., et al. 2010. The developmental transcriptome of *Drosophila melanogaster*. *Nature* 471: 473–479; Gąsiorowski, L., and Hejnol, A. 2019. Hox gene expression in postmetamorphic juveniles of the brachiopod *Terebratalia transversa*. *EvoDevo* 10: 1.

On morphostasis: Wagner, G. P., and Misof, B. Y. (1993). How can a character be developmentally constrained despite variation in developmental pathways? *Journal of Evolutionary Biology* 6: 449–455.

On pattern formation: Waddington, C. H. (1956). *Principles of Embryology*. New York: MacMillan; Meinhardt, H. (1982). *Models of Biological Pattern Formation*. London: Academic Press; Prusinkiewicz, P., and Lindenmayer, A. (1990). *The Algorithmic Beauty of Plants*. New York: Springer.

On the similarities between embryonic development and regeneration processes in planarians: Cardona, A., Hartenstein, V., and Romero, R. (2005). The embryonic development of the triclad *Schmidtea polychroa*. *Development Genes and Evolution* 215: 109–131.

On developmental modules: von Dassow, G., and Munro, E. M. (1999). Modularity in animal development and evolution: Elements of a conceptual framework for EvoDevo. *Journal of Experimental Zoology (Molecular and Developmental Evolution)* 285: 307–325; Schlosser, G., and Wagner, G. P. (eds.) (2003). *Modularity in Development and Evolution*. Chicago: University of Chicago Press; Klingenberg, C. P. (2005). Developmental constraints, modules and evolvability. In B. Hallgrímsson and B. K. Hall (eds.) *Variation: A Central Concept in Biology*. Burlington, MA: Elsevier, pp. 219–247.

On developing organisms as multigenomic systems: Dupré, J. (2010). The polygenomic organism. *The Sociological Review*, 58 (Supplement 1): 19–31; Gilbert, S. F., Sapp, J., and Tauber, A. I. (2012). A symbiotic view of life: We have never been individuals. *The Quarterly Review of Biology* 87: 325–341; McFall-Ngai, M., Hadfield, M. G., Bosch, T. C. G., et al. (2013). Animals in a bacterial world, a new imperative for the life sciences. *Proceedings of the National Academy of Sciences of the United States of America* 110: 3229–3236; Bosch, T. C. G., and McFall-Ngai, M. J. (2011). Metaorganisms as the new frontier. *Zoology* 114: 185–190; Gilbert, S. F., and Epel, D. (2015).

Ecological Developmental Biology: The Environmental Regulation of Development, Health, and Evolution. Sunderland, MA: Sinauer; Bull, M. J., and Plummer, N. T. (2014). The human gut microbiome in health and disease. *Integrative Medicine (Encinitas)* 13: 17–22.

Chapter 7

On the 'inelegant' production of serial structures in fruit flies and leeches: Akam, M. (1989). Making stripes inelegantly. *Nature* 341: 282–283; Weisblat, D. A., and Kuo, D.-H. (2018) Developmental biology of the leech *Helobdella*. *International Journal of Developmental Biology* 58: 429–443.

On the development of an animal's main body axis: Slack, M. et al. (1993), cit.; Minelli, A. (2005). A morphologist's perspectives on terminal growth and segmentation. *Evolution and Development* 7: 568–573.

On the neural crest: Hall, B. K. (1998). Germ layers and the germ-layer theory revisited: Primary and secondary germ layers, neural crest as a fourth germ layer, homology, demise of the germ-layer theory. *Evolutionary Biology* 30: 121–186.

On growth and polarity of fungal hyphae: Schmieder, S. S., Stanley, C. E., Rzepiela, A., et al. (2019). Bidirectional propagation of signals and nutrients in fungal networks via specialized hyphae. *Current Biology* 29: 217–228.

On axon growth: Tessier-Lavigne, M., and Goodman, C. S. (1996). The molecular biology of axon guidance. *Science* 274: 1123–1133.

On left–right asymmetry in snails: Grande, C., and Patel, N. H. (2009). Nodal signalling is involved in left–right asymmetry in snails. *Nature* 457: 1007–1011.

On autotomy not followed by regeneration: Emberts, Z., St. Mary, C. M., and Miller, C. W. (2016). Coreidae (Insecta: Hemiptera) limb loss and autotomy. *Annals of the Entomological Society of America* 109: 678–683.

On asymmetry: Palmer, A. R. (2016). What determines direction of asymmetry: Genes, environment or chance? *Philosophical Transactions of the Royal Society B* 371: 20150417.

On deer antlers and vertebrate teeth: Goss, R. J. (1983). *Deer Antlers: Regeneration, Function and Evolution*. New York: Academic Press; Hall, B. K. (2005). *Bones and Cartilage: Developmental and Evolutionary Skeletal Biology*.

London: Elsevier/Academic Press; Berkovitz, B. K., and Shellis, R. P. (2017). *The Teeth of Non-Mammalian Vertebrates*. London: Elsevier/Academic Press.

On fractals and paramorphism: M'Cosh, J. (1851). Some remarks on the plant morphologically considered. *Transactions of the Botanical Society* 4: 127–132; Arber, A. (1950). *The Natural Philosophy of Plant Form*. Cambridge: Cambridge University Press; Mandelbrot, B. B. (1982). *The Fractal Geometry of Nature*. San Francisco: Freeman; Minelli, A. (2003). The origin and evolution of appendages. *International Journal of Developmental Biology* 47: 573–581; Glasby, C. J., Schroeder, P. C., and Aguado, M. T. (2012) Branching out: A remarkable new branching syllid (Annelida) living in a *Petrosia* sponge (Porifera: Demospongiae). *Zoological Journal of the Linnean Society* 164: 481–497.

Chapter 8

On developmental robustness: Bateson, P., and Gluckman, P. (2011). *Plasticity, Robustness, Development and Evolution*. Cambridge: Cambridge University Press.

On reversible versus irreversible developmental switches: Greene, E. (1989). Diet-induced developmental polymorphism in a caterpillar. *Science* 243: 643–646; Sunobe, T., and Nakazono, A. (1993). Sex change in both directions by alternation of social dominance in *Trimma okinawae*. *Ethology* 94: 339–345.

On phenotypic plasticity: Fusco, G. and Minelli, A. (2010). Phenotypic plasticity in development and evolution. *Philosophical Transactions of the Royal Society of London B* 365: 547–556.

On periodical cicadas: Williams, K. S., and Simon, C. (1995). The ecology, behavior, and evolution of periodical cicadas. *Annual Review of Entomology* 40: 269–295.

On plant temporal phenotypes: Klimešová, J., Martínková, J., and Kočvarová, M. (2004). Biological flora of Central Europe: *Rorippa palustris* (L.) Besser. *Flora* 199: 453–463.

On the periodization of arthropod development: Minelli, A., Brena, C., Deflorian, G., Maruzzo, D., and Fusco, G. (2006). From embryo to adult – beyond the conventional periodization of arthropod development. *Development, Genes and Evolution* 216: 373–383.

Examples of standard tables of development: Theiler, K. (1989). *The House Mouse: Atlas of Mouse Development*. New York: Springer; Campos-Ortega, J. A., and Hartenstein, V. (1985). *The Embryonic Development of* Drosophila melanogaster. Berlin: Springer.

Quotation on p. 146 from p. 111 of Ptolemy, C. (1822). *Ptolemy's Tetrabiblos, Or Quadripartite*, trans. by J. M. Ashmand. Davis and Dickson, London.

On embryonic diapause: Renfree, M. B., and Fenelon, J. C. (2017). The enigma of embryonic diapause. *Development* 144: 3199–3210.

On the Dauer larva of *Caenorhabditis elegans*: Blaxter, M. (2011) Nematodes: The worm and its relatives. *PLoS Biology* 9: e1001050.

On the contrasting age of male versus female body parts in *Hemioniscus*: Goudeau, M. (1977). Contribution à la biologie d'un crustacé parasite: *Hemioniscus balani* Buchholz, isopode epicaride. Nutrition, mues et croissance de la femelle et des embryons. *Cahiers de Biologie Marine* 18: 201–242.

On senescence: Fahy, G. M. (2010). Precedents for the biological control of aging: experimental postponement, prevention, and reversal of aging processes. In Fahy, G. M. (ed.) *The Future of Aging. Pathways to Human Life Extension*. Dordrecht: Springer, pp. 127–225; Shefferson, R. P., Jones, O. R. and Salguero-Gómez, R. (2017). *The Evolution of Senescence in the Tree of Life*. Cambridge: Cambridge University Press; Fusco, G., and Minelli, A. (2019). *The Biology of Reproduction*. Cambridge: Cambridge University Press.

On *Dioscorea pyrenaica*: García, M. B., Dahlgren, J. P., and Ehrlén, J. (2011). No evidence of senescence in a 300-year-old mountain herb. *Journal of Ecology* 99: 1424–1430.

Concluding Remarks

Quoted sentences from Bateson, W. (1908). *The Methods and Scope of Genetics. An Inaugural Lecture delivered 23 October 1908*. Cambridge: Cambridge University Press.

Index